ENVIRONMENTAL SCIENCE, ENGINEERING AND TECHNOLOGY

FOREST ECOSYSTEMS

BIODIVERSITY, MANAGEMENT AND CONSERVATION

ENVIRONMENTAL SCIENCE, ENGINEERING AND TECHNOLOGY

Additional books in this series can be found on Nova's website under the Series tab.

Additional e-books in this series can be found on Nova's website under the e-book tab.

ENVIRONMENTAL SCIENCE, ENGINEERING AND TECHNOLOGY

FOREST ECOSYSTEMS

BIODIVERSITY, MANAGEMENT AND CONSERVATION

NOEL C. ROBERTS
EDITOR

nova publishers
New York

For permission to use material from this book please contact us:
Telephone 631-231-7269; Fax 631-231-8175
Web Site: http://www.novapublishers.com

Library of Congress Cataloging-in-Publication Data

Forest ecosystems : biodiversity, management, and conservation / editor, Noel C. Roberts.
pages cm. -- (Environmental science, engineering and technology)
Includes bibliographical references and index.
ISBN: 978-1-63117-815-3 (hardcover)
1. Forest ecology. 2. Biotic communities. I. Roberts, Noel C., editor of compilation.
QH541.5.F6F667 2014
577.3--dc23

2014013784

Published by Nova Science Publishers, Inc. † New York

CONTENTS

PREFACE

Temperate forests covering 34% of the Earth's land have experienced human disturbances since ancient times and this is where the major industrial and agricultural enterprises are located nowadays. The heaviest human stress has been on European and North American ecosystems. Forest ecosystems are a major control of moisture circulation in watersheds. Boreal forests have been long recognized to influence global moisture circulation considerably. Therefore, it is logical to expect that further large-scale forest cover disturbances, such as catastrophic fires, large-scale insect outbreaks and forest logging, will affect the processes of water resource development and hydrological regime in any river basin. This book discusses the management and conservation of forest ecosystems in several areas in the world which include the Neotropics, Norway, United Kingdom and Siberia.

Chapter 1 – The Neotropic rainforest in Central and South America has some of the highest levels of biodiversity on earth. Its future is impacted, however, by human disturbances usually clearing land for agriculture. Studies are needed to examine those "old fields" after abandonment in order to make wise decisions about rainforest conservation and management. Towards that end the author set up and sampled permanent plots after use as pasture and after Banana, Coffee and Sugarcane cultivation. He also performed seed and seedling experiments in those same fields in order to understand the causes of patterns seen in the plot data. The author focused on sampling the seed rain and on seed mechanisms (predation, pathogens), seed tolerances (germination), seedling mechanisms (herbivory, pathogens) and seedling tolerances (competition). He relates those studies conducted for 17 years in Ecuador and for 10 years in Puerto Rico, and compare those secondary forests with the primary forest nearby.

Chapter 2 – Research dealing with modeling of vegetation dynamics and spatial pattern in a declined mountain spruce forest is closely related to remote sensing, which can discover changes in land cover. Assessing land cover by classification of archived images obtained from a series of Landsat sensors can provide information about the succession of a spruce forest. New approaches can be implemented on the basis of remote sensing, geographic information systems (GISs) and dynamic modeling. This offers a wide range of spatial and temporal analyses to explore spruce forest succession. The described research shows remote sensing data from Landsat sensors including: Thematic Mapper (TM), Enhanced Thematic Mapper Plus (ETM+) and Operational Land Imager (OLI) of Landsat 5, Landsat 7 and Landsat 8, respectively. The area of interest is located in the Tatra National Park, which is part of the Carpathian Mountains in Europe. The explored period from July 1994 to June 2010 is based on available archive images from Landsat 5 TM. An extension of the period to September 2013 can be obtained from archive images from Landsat 7 ETM+ and Landsat 8 OLI. The most important period was the winter season in November 2004 when the territory was swept by gale-force winds with gusts of almost 200 km/hour. In order to explore land cover changes for modelling vegetation dynamics and spatial patterns, unsupervised classification and supervised classification were used to detect land cover classes from image compositions based on Landsat 5 TM bands for the images captured in July 1994 and in June 2010. The target data structures for modelling vegetation dynamics and spatial pattern are formed by a grid of cells that represent each local site. The numerical solution of a spatio-temporal model for dominant seedlings of Picea abies is based on a set of differential equations for the grid of cells. The growth predictions of individual trees are based on a set of algebraic equations. The proposal for a spatio-temporal model in GIS enables other processing to provide prediction of spruce forest regeneration and succession after the winds in November 2004. GIS can integrate all the spatial data together with model predictions and support the decision-making processes in risk assessment analysis and natural conservation.

Chapter 3 – Spruce dominated forests undoubtedly represent one of the conservation ironies of the 21st century in the UK. These introduced tree species are planted extensively in commercial forestry and now offer a refuge to the endangered native Eurasian red squirrel (Sciurus vulgaris) in the face of competition from the introduced North American grey squirrel (S. carolinensis). The present chapter provides a concise historical overview of red squirrel population decline within the UK, describes the respective ecology and competitive interactions of both red and grey squirrels and the subsequent evolution of current conservation strategies. The importance of spruce dominated forests in the conservation of red squirrels is critically reviewed; population dynamics and habitat use are explored with particular focus upon patterns of tree seed availability, and the integration of squirrel conservation with parallel commercial forest management targets are described. Knowledge gaps in the proactive management of remnant red squirrel populations are presented, and we explore the potential impact of emerging disease threats to both squirrels and UK forest ecosystems.

Chapter 4 – The heat- and water vapor transfer in wood and the closely linked wood structural properties have been studied extensively using steady state conditions. The experimentation on Norway spruce confirmed significant impact of surface morphology on established boundary conditions and there from on changed heat in mass transfer. The real bearing wood surface profiles actually often follow Abbot curve with a signed increase of bearing surface towards the inner wood tissue. This leads to the ascertaining of the inverse proportional relation between the contact thermal- and surface mass transfer resistance and the contact surface of wood. The results confirm the existence of contact thermal- and surface mass resistance when using the hot plate method or the steady state diffusion Cup method. Findings support discussion on the credibility and careful use of steady state experimentation on heat- and mass transfer at Norway spruce wood.

Chapter 5 – Elucidating the relationship between plant diversity and soil erosion on different scales can give insights into the interactions between botanical processes and the earth's surface, and is crucial for the establishment of theories on biodiversity and ecosystem functions. In Yunnan province, China, we studied local ecosystems that showed gradual changes in plant diversity and ecosystem maturity, watershed ecosystems that only presented gradual changes in ecosystem maturity, and 11 floral subregions, representing regional ecosystems, in which plant diversity was assessed based on the recorded statistics of 13,850 seed plants. Runoff plots and weir method were used to determine soil erosion in the local and watershed ecosystems from 2000 to 2006. Soil erosion of the regional ecosystems was determined by superposing about 10,000 patches of soil-erosion maps, with predetermined erosion grades, on the 12 floral subregions based on geographical information system (GIS). Surface runoff and soil erosion decreased with increasing plant diversity and ecosystem maturity in local ecosystems, but showed obvious differences under the conditions of similar plant diversity and different levels of ecosystem maturity in the watershed ecosystems. Soil erosion declined with increasing diversity in regional ecosystems. Soil erosion control was gradually enhanced with increasing diversity and ecosystem maturity and the function was mainly dependent on ecosystem maturity. Increasing diversity leads to higher plant cover and density in local ecosystems, thereby promoting erosion control. Covering, diffusion and density effects explain the enhanced process. Ecosystem maturity played a greater role in soil erosion control than plant diversity, which presented a weaker effect on the function. An enhanced function of soil erosion control with increasing diversity predicts an increase in ecosystem productivity. This study provides an integrated understanding of the relationships between diversity and ecosystem functions on different scales.

Chapter 6 – Siberian rivers account for a great part of water discharge into the Arctic Ocean. Along with the regional climate and the presence permafrost, hydrological processes occurring in catchments of the boreal taiga rivers are controlled by forest harvesting levels. Forests of the Angara region and West Sayan have been harvested extensively over the past fifty years. As a result, forest area has been reduced and forest age structure and composition have changed. These changes are reflected in the natural water budget structure and hydrological regimes of areas. The study revealed that hydrological effects of forest logging conducted in North Angara region characterized by a highly continental climate and, hence, severe forest growing conditions differ from those in West Sayan.

In: Forest Ecosystems
Editor: Noel C. Roberts
ISBN: 978-1-63117-815-3

Chapter 1

PRIMARY VS. SECONDARY FORESTS IN THE NEOTROPICS: TWO CASE STUDIES AFTER AGRICULTURE

Randall W. Myster*****
Biology Department, Oklahoma State University, Oklahoma City, US

ABSTRACT

The Neotropic rainforest in Central and South America has some of the highest levels of biodiversity on earth. Its future is impacted, however, by human disturbances usually clearing land for agriculture. Studies are needed to examine those "old fields" after abandonment in order to make wise decisions about rainforest conservation and management. Towards that end I set up and sampled permanent plots after use as pasture and after Banana, Coffee and Sugarcane cultivation. I also performed seed and seedling experiments in those same fields in order to understand the causes of patterns seen in the plot data. I focused on sampling the seed rain and on seed mechanisms (predation, pathogens), seed tolerances (germination), seedling mechanisms (herbivory, pathogens) and seedling tolerances (competition). Here I relate those studies conducted for 17 years in Ecuador and for 10 years in Puerto Rico, and compare those secondary forests with the primary forest nearby.

---

* Email: myster@okstate.edu, personal web site: https://sites.google.com/site/rwmyster.

1. Introduction

The Neotropics is defined as the American landmass and associated islands between the Tropic of Cancer (~23.5° N latitude) and the Tropic of Capricorn (~23.5° S latitude). These two major delineations of the earth mark the most northerly and southerly positions at which the Sun may appear directly overhead at its zenith. This greatly reduces light and temperature seasonally, thus allowing year-round plant growth. Within the Neotropics, temperatures can attain 38° C but have a usual range between 17° C and 27° C, and annual precipitation can reach 6 m with seasonal wet and dry cycles (Mabberley, 1992). That rainfall is influenced by El Nino southern oscillations that occur every 3-7 years and can cause floods, droughts, and fires (Sanford et al., 1985). Many Neotropical soils are old, highly weathered, leached oxisols and ultisols, well drained, and have a thin litter layer (Mabberley, 1992; Finegan, 1996) with low organic matter content. Also common in the Neotropics are andisols derived from volcanic activity, which are very fertile and support intensive agriculture. Many Neotropical soils have a higher clay content than do temperate soils and are low in phosphorus. Consequently, plant growth in the Neotropics may be more limited by phosphorus than by other nutrients (Vitousek, 1984).

The Neotropics includes more tropical rainforest than any other place and is one of the most important reserves of biodiversity on Earth. In addition to plants and animals, the Neotropics is home to a large amount of the biodiversity of indigenous peoples as well: their cultures, languages and dialects (some 350–400 are left). Before the arrive of humans, the Neotropics was dominated by primary broadleaf evergreen forests, both moist and dry, flooded and unflooded . In addition grasslands, savannas and shrublands were common at both high and low elevations (see chapters in Myster 2012c), and coniferous forests, deserts and mangroves were also found (Schultz 2005). Whereas all biomes in the Neotropics could be cleared for cultivation, forests are most often used for cultivation. These forests have high plant diversity where mycorrhizae are common, effectively helping plants forage for phosphorus and other soil nutrients. Deserts are typically nonarable, tropical grasslands usually have periodic droughts, and large areas are often under water, making cultivation much more difficult in these areas (Mabberley, 1992).

The first humans that arrived in the Neotropics cultivated crops. Whatever the technique they used, those fields were at some point abandoned and began to recovery back to the primary forest they once were, creating large areas of secondary forest, whose conservation and management are key to the future of the Neotropics. These post-agricultural areas – the process was often called old field succession – have been studied for some decades and have helped ecologists gain insights into the processes that structure plant communities, such as the role of history and initial conditions in community development (Myster and Pickett 1990b, 1994). Examination of old fields has also led to the rise of important ecological theories, such as the initial floristic composition hypothesis (Egler 1954) and the resource ratio hypothesis (Tilman 1988).

Using permanent plots (Myster and Malahy 2008) established at abandonment as the backbone of old field research (e.g., in New Jersey USA [Buell et al. 1971, Myster and Pickett 1990a]), scientists have been successful in finding the pattern of response over time after abandonment from crops (Myster and Pickett 1992) and in discovering many of the mechanisms and tolerances that determine that pattern (Myster 1993, Bazzaz 1996). Furthermore, studies of post-agricultural succession have provided a framework for comparison and evaluation of various theories of succession and community development (Finegan 1984).

Whereas a considerable amount of old field and pasture research has taken place in the temperate regions of the United States (Oosting 1942, Miles 1979, Bazzaz 1968, Buell et al. 1971, Pickett 1982, Castro et al. 1986, Tilman 1988, Myster 1993) and Europe (Miles 1979, Castro et al. 1986), the recovery of agricultural areas is vital to ecosystems all over the world (Borhidi 1988.) where they can form ecotones between forest and grasslands (Myster 2012c). Particularly in the Tropics, the cutting down of wooded areas (Grainger 1988, Fearnside 1993) for agriculture and pasture (Skole and Tucker 1993) is the major cause of deforestation (Fearnside 1993, Buschbacher 1986, Uhl et al. 1988, Thomlinson et al. 1996).

Consequently, recovery of Neotropical areas after agriculture (Donfack et al. 1995, Fernandes and Sanford 1995, Quintana-Ascencio et al. 1996) is relevant to such important issues as forest regeneration (Grainger 1988, Skole and Tucker 1993, Brown and Lugo 1990, Myers 1991, Singh 1993), forest ecosystem restoration and conservation (Brown and Lugo 1990, Lugo 1992), sustainability and management of agriculture (Brown and Lugo 1990, Serrao and Toledo 1990), maintenance of biodiversity (Brown and Lugo 1990, Aide and Cavelier 1994), and impacts of global climate change on forest dynamics (Padoch and Vayda 1983, Hobbie 1992, Mabberley 1992, Keller et al. 1993). Neotropical areas recovering from agriculture may also serve as a buffer between primary forest and more intensely human-influenced areas (Brown and Lugo 1990). Finally with future increase of global warming, and other human induced effects, forest disturbances like old field succession should affect larger and larger parts of the earth (Bazzaz 1996).

The nature of the recovery of old fields depends on their disturbance regime, which is characterized by severity, size, spatial location, and frequency (Myster 2007b). The measurement of severity may be best accomplished by sampling the loss of plant biomass and determining whether such loss includes loss of plant meristems (Raunkaier 1934, Pickett and White 1985, Tilman 1988). Severity of old fields is moderate, compared to the more severe landslides (Myster and Walker 1997, Myster and Sarmiento 1998) and the less severe treefall (Everham et al. 1996) because old fields maintain an intact soil profile, roots, and humus layer despite the removal of above-ground vegetation. In size, old fields do not usually exceed a few tens of hectares (they are larger, however, than many Neotropic treefall gaps: Brokaw 1982), and crops that are planted in them often require specific locations (e.g., Coffee grows well in the mountains while Banana and Sugarcane do best in the valleys).

The frequency of an old field's disturbance regime is complex and includes the number of times a field has been cultivated, the duration of each cropping period, the time between fallow periods, and the order of crop rotation. Usually fields are fallow longer than they are in crop and can be either reused with different crops and fallow periods until abandoned or used for 5-10 years until productivity declines (Staver 1990). As the length of time an area is in crop increases, the length of successional, or recovery, time may also lengthen. For example, a short cropping period can lead to large contributions from the seed bank and trees may regenerate quickly from stump or root sprouts, whereas an increased cropping period brings about the decay of rootstocks and tree stumps, thus providing for grass invasion and prolonged tree invasion.

Among other important abandonment conditions that influence old field recovery are the historical effects of the past crop (a crop's "signature": Myster and Pickett 1988, Myster and Pickett 1990b, Myster and Pickett 1994), the year of abandonment, the season of abandonment, the percent of border with forest (Bazzaz 1968), and a field's plowing record (whether the field was plowed under or left fallow when abandoned: Myster and Pickett 1990a). In particular, studies of post-agricultural succession in the temperate zone have demonstrated the importance of a crop's signature in determining old field patterns (e.g., alteration of successional pathways and species composition and abundance for up to 8 years after abandonment: Myster and Pickett 1988, Myster and Pickett 1990b, Myster and Pickett 1994) where the length of a crop's persistence after abandonment (and other lingering effects) may be related to its relative amount of root biomass (author, pers. obs.).

In this chapter I relate my studies in abandoned fields of Ecuador and Puerto Rico after they were used to grow common Neotropic crops, such as Sugarcane (*Saccharum officinarum*), Banana (*Musa* spp.), and Coffee (*Coffea* spp.), and after use as pasture where pasture grasses usually establish themselves naturally after cropping but can also be purposely planted (e.g., *Setaria sphacelata* in Ecuador). I assume pastures are a kind of old field succession for several reasons. First, they are usually converted from depleted agricultural fields when cows are allowed to enter. Second, they may continue to have remnants of the previous past crops growing in them for some years. And third, they can be planted (cropped) with exotic grasses or be colonized by local grass species due to cattle activity such as trampling of vegetation, creation of hummocks, deposition of dung, and soil compaction.

For the studies reported here, all fields were first cleared from forest using shifting or "slash and burn" (Uhl 1987) agriculture, which occurs when natural vegetation is first cut down to the ground and then burnt. Although burning produces a short pulse of nutrients like phosphorus, it generally volatilizes both organic matter and nitrogen while raising soil temperature and, in the long term, makes soil poor in both nutrients and the seed bank. Shifting agriculture also entails a rotation of fields rather than crops that is accompanied by long fallow periods. The decision as to when to rotate fields and for how long to leave them fallow depends on such factors as weed and insect loads in the fields, soil fertility, production, labor availability, and local dietary needs (Staver 1980). This forest-clearing method causes carbon loss, a mixing of the soil horizons, and a higher degree of soil aeration (Bazzaz 1996). Shifting agriculture also lowers the level of advanced regeneration from sapling banks and from root/stem resprouting so that seed and seedling dynamics (Grubb 1977) dominate regeneration after agriculture.

Old field recovery goes through several structural and conceptual stages: (1) domination by past crop and its various effects, (2) patch dynamics of grass, past crop remnants, asexual shrubs, and other plants, (3) tree invasion, and (4) development of a closed-canopy secondary forest. After a closed canopy has developed, old field plants may eventually enter a thinning state when dead plants in the canopy are replaced not with new individuals but with modified growth of existing plants. There is a physiological limit to this kind of growth, however, and at some point every plant will die and be replaced.

Indeed plants are always involved in the on-going dynamic process of plant-plant replacement (Busing 1996, Myster 2012b). Plant-plant replacement does not necessarily, however, involve two plants only for more than one plant can be replaced by just one or only one plant can be replaced by more than one. As plant-plant replacements proceed, the two spaces each plant defines (its phyto-space defined by its biomass and necromass, and its surrounding neighborhood space) can be occupied and reoccupied. Because plant replacements occur slowly, permanent plots are needed in order to sample the dynamic replacements all plants are engaged in. Replacements create changes in plant abundances, which may also create changes in plant composition that lead to the emergence of other patterns at larger spatial and temporal scales (e.g., successional rate and direction, patch dynamics, plant distribution, exotic plant invasion, plant associations).

Care must be taken, however, with the observation of these plant patterns. For example, even though individual plants continue to be replaced, preservation of species abundances and/or species composition at a larger scale of organization may still occur, giving the impression of a "coexistence" of species (when the identity or number of plant species remain the same despite changes in plant abundances) or of a plant community at "equilibrium" (when neither plant species nor plant abundances change). Existence of high-level patterns that do not change very much over time should not obscure the fact that such patterns are the product of a dynamic and ever changing process of plant-plant replacement. Any plant pattern is only a snapshot of the vegetation at a given instance in time and at a specific scale of observation, whereas plant-plant replacements occur continuously.

Replacements are controlled by species tolerances and mechanisms which affect plant growth and the definition of plant spaces. Tolerances include germination, growth, seedling competition while mechanisms encompass dispersal, seed pool, seed predation, seed pathogens, seed germination, seedling predation, seedling pathogens and seedling herbivory (Myster and McCarthy 1989, Myster and Pickett 1993, Myster 2004a, Myster 2004b, Myster 2007a). The process of plant-plant replacement is at the heart of the dynamic nature of terrestrial plant communities both over successional time and over space (e.g., at the edges of plant distributions where gradients have a major effect on plants: Whittaker 1975). It seems that plants wait not to be counted (sensu Harper 1977) but to be replaced! Thus the ultimate reality of plant communities are the plant-plant replacements, for they produce vegetation patterns at all larger temporal and spatial scales. The underlying process of the two forests discussed in this paper, as for all plant communities, is plant-plant replacement (Myster 2012b) which produced all plant population and community patterns. These permanent plot studies, and others like them, provide baseline data on forest dynamics and fluctuations of forest structure. This knowledge will enable conservationists to develop sound management techniques for these forests in order to better utilize them as societal and human needs arise in the future.

2. Case Study: Maquipucuna Reserve Cloud Forest, Ecuador

Maquipucuna Reserve (0°05'N, 78°37'W; www.maqui.org; Sarimento, 1997; Rhoades et al., 1998; Rhoades and Coleman, 1999) lies between 1200 m and 1800 m and is classified as tropical lower montane cloud forest (Edmister, 1970). It has deeply dissected drainages with steep slopes and receives between 2 m and 5 m of rainfall a year. The temperature ranges between 14°C and 25°C, with an average temperature of 18°C. The reserve's fertile andisol soil is developed from volcanic ash deposits. Maquipucuna Reserve is located within the Choco-Andean corridor which is one of the top five biodiversity hotspots on earth (Sarimento, 1995).

2.1. Maquipucuna Reserve Cloud Forest Floristics and Physical Structure

In May 2012, my field assistants and I set up a 1 ha (50 m x 200 m) plot in a primary Cloud forest at Maquipucuna Reserve. We tagged and measured the diameter at breast height (dbh) of all trees at least 10 cm dbh in each of the 100 10 m x 10 m continuous subplots that make up the 1 ha plot. The dbh measurement was taken at the nearest lower point where the stem was cylindrical and for buttressed trees it was taken above the buttresses. The tagged trees were identified to species, or to genus in a few cases, using Romoleroux et al. (1997) and Gentry (1993) as taxonomic sources. We also consulted the web site of the Missouri Botanical Garden <www.mobot.org>.

From this data set, I first compiled Floristic tables of family, genus and species. Then I generated these structural parameters (1) the total number of stems in the 1 ha plot, the mean and maximum among those stems, and the total number of stems divided into four size classes: 10 < 20 dbh, 20 < 30 dbh, 30 < 40 dbh and ≥ 40 dbh, (2) the stem dispersion pattern (random, uniform, clumped) computed by comparing plot data to Poisson and negative binomial distributions using Chi-square analysis and, if clumped, greens index was also computed to access degree of clumping (Ludwig and Reynolds, 1988; Myster and Pickett, 1992), (3) canopy closure using the formula in Buchholz et al. (2004) for tropical trees with the resulting percentage of the 1 ha plot area closed, (4) total basal area as the sum of the basal areas of all individual stems (Πr^2; where r = the dbh of the individual stem / 2) and (5) above-ground biomass (AGB) using the formula in Nascimento and Laurance (2001) suggested for tropical trees of these stem sizes.

I found (author, unpub. data) there were a total of 18 families in the 1 ha cloud forest study plot (Table 1). Lauraceae was by far the most common family and all families had a low number of genera and species (dominated by very few genera and species).

The families Actinidaceae, Fabaceae and Verbenaceae were also common, but there were three families (Asteraceae, Brunneliaceae, Monimiaceae) with only one stem.

The families Asteraceae, Brunneliaceae, Fabaceae, Melastomataceae, Moraceae, Myrtaceae, Piperaceae, Rubiaceae, Solaneceae, Tiliaceae, Urticaceae and Verbenaceae had a monotonic decline in stem number as stems get thicker (Table 2). This was not true, however, of the families Actinidaceae, Cecropiacea, Lauraceae, Mimosaceae, Monimiaceae and Myristicacea. The most common species were *Erythrina megistophylla* and *Nectandra acutifolia* (Table 3).

The 1 ha study plot also had a reverse J stem size distribution pattern with less than half of the stems found in some lower elevation *terra firme* plots which were reflected in all size classes (Table 4). Stems in the study plot were a bit larger than the *terra firme* forests but species richness was very much lower (Table 4) which lead to less basal area, biomass and canopy closure (~46%: Table 4). Some of these trends continued – less stems, ticker stems – when we compare the study plot with lower elevation palm forest and white sand forest (Table 5) but the study plot was more closed than these two forests. Basal area and biomass were similar among these forests (Table 5).

Finally, there were a reasonable number of plant families represented in the study plot but a low number of genera and species per family. Stem number decreased with increasing stem size within individual families about as often as it did not. There were large numbers of dominate species found. Going up the mountain from lowland *terra firme* reduces the number of stems at least by half, but size of stems increases to a small degree. With most of the structural parameters cloud forest is between *terra firme* and palm/white sand forest.

Table 1. Each family sampled in the 1 ha cloud forest study plot sorted in decreasing alphabetical order with the total number of stems in parenthesis followed by the total number of genera and the total number of species

Family (total number of stems)	number of genera	number of species
Actinidaceae (24)	1	1
Asteraceae (1)	1	1
Brunneliaceae (1)	1	1
Cecropiacea (7)	1	1
Fabaceae (49)	1	1
Lauraceae (118)	1	2
Melastomataceae (10)	2	2
Mimosaceae (7)	1	1
Monimiaceae (1)	1	1
Moraceae (2)	2	2
Myristicacea (9)	2	2
Myrtaceae (4)	1	1
Piperaceae (5)	1	1
Rubiaceae (11)	4	4
Solaneceae (5)	1	1
Tiliaceae (2)	1	1
Urticaceae (12)	1	1
Verbenaceae (24)	1	1

Table 2. The family and total stems for each family from table 1 divided into size classes based on diameter at breast height (dbh) measured in whole cm

Family	10 < 19 cm	20 < 29 cm	30 < 39 cm	40 cm or greater
Actinidaceae (24)	8	13	3	0
Asteraceae (1)	1	0	0	0
Brunneliaceae (1)	1	0	0	0
Cecropiacea (7)	0	4	1	2
Fabaceae (49)	43	5	1	0
Lauraceae (118)	30	39	39	10

Table 2. (Continued)

Family	10 < 19 cm	20 < 29 cm	30 < 39 cm	40 cm or greater
Melastomataceae (10)	10	0	0	0
Mimosaceae (7)	5	2	0	0
Monimiaceae (1)	0	1	0	0
Moraceae (2)	1	1	0	0
Myristicacea (9)	4	6	0	0
Myrtaceae (4)	4	1	0	0
Piperaceae (5)	5	0	0	0
Rubiaceae (11)	9	1	1	0
Solaneceae (5)	5	0	0	0
Tiliaceae (2)	2	0	0	0
Urticaceae (12)	12	0	0	0
Verbenaceae (24)	13	9	1	1

Table 3. All species in the study plot with at least 4 stems, sorted by family, then genus and species

Family	Genus	species	number of stems
Actinidaceae	Saurauia	priannana	4
Cecropiacea	Cecropia	sp.	7
Fabaceae	Erythrina	megistophylla	49
Lauraceae	Nectandra	acutifolia	118
Melastomataceae	Miconia	sp.	10
Mimosaceae	Inga	sp.	7
Myristicacea	Otoba	gordoniifolia	10
Myrtaceae	Eugenia	sp.	5
Piperaceae	Piper	sp.	5
Rubiaceae	Policourea	sp.	11
Solaneceae	Cestrum	sp.	5
Urticaceae	Urera	sp.	12
Verbenaceae	Aegyphila	sp.	24

Table 4. Physical structure parameters for all trees at least 10 cm dbh sampled in the 1 ha cloud forest study plot, a 1 ha *terra firme* plot (author, unpub. data sampled in western Peru) and in three other *terra firme* forest plots (all sampled in Ecuador). All data are expressed per hectare, regardless of how they were sampled

Parameter	cloud forest study plot	*terra firme*	*terra firme*[1]	*terra firme*[2]	*terra firme* plot[3]
Stem density :					
Total	294	519	604 – 725	605.3	734
10 < 19 dbh	153	288	--	--	--
20 < 29 dbh	82	121	--	--	--
30 < 39 dbh	46	50	--	--	--
40 or greater	13	60	--	--	--
mean	23.2	20.8	--	21.7	--
Family richness	18	--	--	--	--
Genus richness	24	--	--	--	--
Species richness	25	302	--	--	--
Basal area (m^2)	11.2	17.6	22.2 – 31.2	32.6	22.2
Above-ground					
biomass (Mg)	198.4	334.2	--	457.8	--
Canopy:					
Closure (m^2)	4633.4	6921.4	--	--	--
per ha (%)	46.334	69.214	--	--	--

[1]Valencia et al. 2004abc, [2]Haugaasen and Peres 2006, [3] Korning et al. 1990.

Table 5. Physical structure parameters for all trees at least 10 cm dbh sampled in the 1 ha cloud forest study plot, in a 1 ha palm forest (author, unpub. data) plot and in a 1 ha white sand forest (author, unpub. data) plot. All plots were sampled at the study site. All data are expressed per hectare, regardless of how they were sampled

Parameter	cloud forest study plot	palm plot	white sand plot
Stem density:			
Total	294	449	403
10 < 19 dbh	153	253	241
20 < 29 dbh	82	176	95
30 < 39 dbh	46	20	63
40 or greater	13	0	3
mean	23.2	14.1	19.8
Family richness	18		
Genus richness	24		
Species richness	25	57	103
Basal area (m^2)	11.2	11.15	12.52
Above-ground			
biomass (Mg)	198.4	267.1	387.7
Canopy:			
Closure (m^2)	4633.4	2356.7	1441.2
per ha (%)	46.334	23.567	14.412

In comparison to the 1 ha study plot, the cloud forest at Monteverde, Costa Rica, sampled for all trees at least 10 cm dbh, had more stems per ha (555 vs. 294) and more basal area per ha (62 m^2 vs. 11.2 m^2) than the study plot (Nadkarni et al., 1995) but it also showed a reverse J size distribution for all stems taken together. There were many families in common between the two plots: Asteraceae, Cecropiacea, Fabaceae, Lauraceae, Melastomataceae, Moraceae, Myristicacea, Myrtaceae, Myrtaceae, Piperaceae, Rubiaceae, Solaneceae, Tiliaceae, Urticaceae and Verbenaceae (Table 1) and Fabaceae and Lauraceae were most common. Cecropia was the only genus in common.

At other cloud forests in the Neotropics, stem density (for all trees at least 10 cm dbh) was also high (Costa Rica, 553 per ha: Heaney and Proctor, 1990), but low in Jamaica (52 per ha: Tanner, 1977) and Ecuador (46 per ha: Grubb et al., 1963), which puts the study plot somewhere in between. In Puerto Rican "dwarf" cloud forest there were many small stems (3671 per ha) but the basal are was comparable to Monteverde in Costa Rica (49.1 m^2 per ha: Weaver et al., 1986). In a cloud forest closer to Ecuador (Venezuela) sampled to a dbh of at least 10 cm, the stem density (365.9 – 850.1 per ha) and species richness (9.13 – 14.45 per ha) compared better to the study plot than previous reported (Schwarzkopf et al., 2011).

2.2. Regeneration of the Maquipucuna Reserve Cloud Forest After Use as a Pasture and After Sugarcane and Banana Cultivation

Next to the primary cloud forest, there are secondary forests of recovering Banana plantations, Sugarcane plantations, and seeded pastures (Sarmiento 1997, Zahari and Augspurger, 1999). Six of these old fields are part of a long-term study (Myster, 2004a; Myster, 2004b; Myster, 2007a; Myster 2010) of 17 years of continuous sampling. That study is funded by the National Science Foundation (Myster, 2012a) as part of the long-term ecological research (LTER) program in Puerto Rico (Brokaw et al. 2012). In June of 1996, I started the study by selecting six just abandoned agricultural fields for study: two recent Sugarcane (*Saccharum officinarum*) plantations denoted left and right when facing the forest, two recent Banana (*Musa* sp.) plantations denoted left and right when facing the forest, and two recent pastures seeded in *Setaria sphacelata* denoted left and right when facing the forest. Although native to Africa, this *Setaria* grass is being planted more and more extensively in the Neotropics because it can withstand heavy grazing by spreading asexually and can form large crowns. Well accepted by cattle, the *Setaria* grass is tolerant of relatively cold temperatures and short-term waterlogged conditions, both common at the Maquipucuna Reserve.

In each field, twenty-five 5m × 2m contiguous plots were laid out. All six fields were 250 m^2 rectangles located (1) within a few hundred meters of each other, (2) at the lower elevations of the reserve, and (3) with the 25 m plot border next to primary cloud forest. The plots did not have any remnant trees or sprouting tree roots at the beginning of the study, and their tree seed bank was very small. Starting in 1997, each of the 5 m x 2 m subplots of each of the six plots were sampled annually to identify each of these plant species and (1) for percent cover of each plant species — an indication of a species' ability to capture light and, therefore, to dominate these areas in the process of becoming forested communities — estimated visually in relation to each plot's area, and (2) for diameter at breast height (dbh) of each tree stem at least 1 cm. Maquipucuna plant taxonomists, trained at the University of Georgia, USA, where voucher specimens are kept on file, assisted in the identification of species by using specimens located on site.

Permanent plots on abandoned pastures show life-form changes such as a domination by grass (e.g., *Panicum* sp., *Axonopus compressus*) that peaks in cover in year three and then declines, fem (*Nephrolepsis* sp., *Thelpteris deltoidea*) and herbaceous (e.g., *Commelina* sp., *Desmodium adscendens*, *Clidemia hirta*) cover that also peaks in year three but at lower cover levels than grass, woody vines (e.g., *Ipomea* sp.) that continue to increase in cover over the first five years, and trees and shrubs that enter, peak, and decline in an individualistic pattern. Permanent plots in sugarcane fields show similar trends of grass dominating early (mostly sugarcane itself, *Saccharum officinarum*) and then declining to 50% cover levels, with *Panicum* spp. and *Brachiaria subquadripara* also present, ferns (e.g., *Nephrolepsis* sp., *Thelypteris deltoidea*) and forbs (e.g., *Desmodium* sp., *Bidens pilosa*, *Lantana camara*) as a small part of the flora that begins to decline immediately with cover of woody species increasing steadily, trees showing this gradual increase with members of the family Melastomataceae most common, and *Piper aduncum* more common here than in other fields.

In the plots located in recovering banana fields, grass and *Musa* sp. dominate early and then decline, but to greater cover levels than in the sugarcane. Ferns attain greater cover levels in banana than in sugarcane but forbs are a small part of the flora. Also, woody species increase, but at smaller cover levels than in sugarcane. There were many of the same tree species in banana fields as in sugarcane fields, but with additional species (e.g., *Cecropia monostachyma*) present. Finally, in the seeded pastures, grass dominated almost completely (mainly the planted grass *Setaria sphacelata*), with a few woody plants showing in year five. Fern and forb cover was small, and pastures had few trees, but the species they did have were common to the sugarcane and banana fields.

Analysis of the data showed that in sugarcane fields, grass dominated early (mostly sugarcane itself, *Saccharum officinarum)* and then declined to 50% cover levels by Year 5, with *Panicum* spp. and *Brachiaria subquadripara* also present. The ferns (e.g., *Nephrolepsis* sp., *Thelypteris deltoidea)* and forbs (e.g., *Desmodium* sp., *Bidens pilosa)* were always a small part of the flora and began to decline immediately, but cover of woody species increased steadily. Number of trees also increased, with members of Melastomataceae most common. In sugarcane fields, *Acalypha platyphylla, Piper aduncum,* and *Vernonia pallens* were common. Further, *P. aduncum* was more common in sugarcane fields than in other fields, and there were 18 tree species in sugarcane fields and 21 in banana fields. In banana fields, grass and *Musa* sp. dominated early and then declined but still had greater cover levels than grass in sugarcane after five years. Ferns attained greater cover levels in banana than in sugarcane whereas forbs were always a small part of the flora. Again, woody species cover increased over time. *Piper aduncum* was frequent in the permanent plots. There were many of the same tree species in banana fields and in sugarcane fields, with few additional species (e.g., *Cecropia monostachyma)* present in either field type. In pastures, grass dominated almost totally (mainly the previously sown grass *S. sphacelata)* although a few woody plants showed in Years 4 or 5. Cover of ferns and forbs was small. Pastures had few trees and the species they did have were common to the other fields (Myster 2007a).

Total cover, an indicator of stratification and developing structure, was greater in the pastures than in the other fields. The cover started at 200% in pastures and gradually declined. Species richness was greater in sugarcane fields than in the other fields, with more species in pastures than in banana fields.

The total number of stems was greatest in sugarcane fields and pastures whereas mean stem height was greatest in the pastures. Basal area increased to 500 cm^2 in sugarcane and banana fields but remained low in pastures. Community parameters in pasture plots show after five years (1) two or three strata or horizontal levels that developed quickly, (2) total stems at 200 per 250 m^2 plot (compared with 48-152 stems per 100 m^2 plot in Brazil; Uhl et al. 1988), (3) tree richness constant at 20 species (also 20 in Brazil but in a 100m^2 plot; Uhl et al., 1988), (4) tree-stem evenness approximately 0.7, (5) productivity peaking at year two at 0.5 kg/m^2/year (10 tons per ha per year in Brazilian pastures; Uhl et al. 1988) and then declining slightly, (6) turnover (for the formula, see Myster and Pickett, 1994) decreasing over the first five years to 50% and (7) mean height at 250 cm and total basal area of stems at 10 00 cm^2. For the sugarcane fields, total cover, an indicator of stratification and developing structure, was greater compared with the other fields, and banana greater than pasture, starting at 200% cover and gradually declining with time. Species richness was also greater in the sugarcane fields compared with the other fields, and banana had more species than did pasture. Total stems were slightly more in the banana fields compared with the sugarcane fields, and pastures had few stems. Compared with banana, mean stem height was also greater in the sugarcane and increased through time, while pasture stems remained small. Basal area stabilized at 500 cm^2 for sugarcane and banana fields (Buschbacher et al., 1988).

Table 6. Significant Spearman rank correlation coefficients among all plant species in the *Left Banana Plantation* over the first 10 years of succession. Significant positive associations are indicated by a "+" and significant negative associations are indicated by a "–". Plant species and families are indicated as follows: Acalypha pladichephalus (A), egonia spp. (B), Geonoma undata (C), Cyathea spp. (D), Musa spp. (E), Pilea spp. (F), Anthurium spp. (G), Trichipterix pilosissima (H), Nectandra spp. (I), Ochroma spp. (J), Baccharis spp. (K), Anthurium spp. (L), Setaria spp. (M), Boconia frutescens (N), Piper aduncum (O), Erythrina megistophyla (P), Vernonia patens (Q), Hedyosum spp. (R), Commelina diffusa (S), Althernantera spp. (T), Siparuna piloso-lepidota (U), Solanum spp. (V), Vernonia spp. (W), Digitaria sanguinalis (X), Inga spp. (Y), and Passiflora spp. (Z)

	B	C	D	E	F	G	H	I	J	K	L	M	N	O	P	Q	R	S	T	U	V	W	X	Y	Z
A	+	+	+	+	+	+	+	–	+	+	+	+	–	+	+	+	+	+							
B		+	+	+	+	+	+	+	+	+	+	+	+	+	+	+	+	+	–	+	+	+			
C			+	+	+	+	+	+	+	+	+	+	+	+	+	–	–								
D				+	+	+	+	+	+	–	+	–	+												
E					+	+	+	+	+	+															
F						+	+	+	+	+	+	–	+	+											
G							+	+	+	+	+	+	+	–	+										
H								+	+	+	+	+	+	+	+	+	+	+							
I									+	+	–	+													
J										+	+	+	+	+	–	+	+	+	+	+	+	+	+		
K											+	+	+	+	+	+	+	+							
L												+	+	+	+	+	+								
M													+	+											
N														+											
O															+	+	+	+	–						
P																+	+	+							
Q																	+	+	+	+	+	+			
R																		+	+	+	+				
S																			+	+	+				
T																				+	+				
U																					+	+	+	+	+
V																						+	+		
W																							++	+	+ +

Table 7. Significant Spearman rank correlation coefficients among all plant species in the *Ecuador Right Banana Plantation* over the first 10 years of succession. Significant positive associations are indicated by a "+" and significant negative associations are indicated by a "–". Plant species and families are indicated as follows: *Acalypha pladichephalus* (A), *Costus* spp. (B), *Musa* spp. (C), *Solanum muricatum* (D), Piperaceae (E), *Setaria* spp. (F), *Tagetes terniflora* (G), *Begonia* spp. (H), *Cuphea cartagenensis* (I), Polypodiaceae (J), *Vernonia patens* (K), *Brugmansia* spp. (L), *Digitaria sanguinalis* (M), Urticaceae (N), *Chusquea* spp. (O), *Nectandra* spp. (P), Piperaceae (Q), *Commelina diffusa* (R), *Erythrina megistophyla* (S), *Heliotropium* spp. (T), *Inga* spp. (U), *Musa acuminate* (V), *Chenopodium album* (W), *Crataegus monogyna* (X), *Boconia frutescens* (Y), and *Cecropia monostachyma* (Z)

	B	C	D	E	F	G	H	I	J	K	L	M	N	O	P	Q	R	S	T	U	V	W	X	Y	Z
A	+	–	–	+	+	+	+	+	+	–	+	+	+	–	+										
B		+	–	+	+	+	+	–	+	+	+	+	+	+											
C			+	+	+	+	+	+	+	+	+	+	+												
D				+	+	+	+	+	+	+	+														
E					+	+	+	–	+	+	+	+	+												
F						–	–	+	+	+	+														
G							+	+	+	+	+	+	+	+	+										
H								+	+	+	+	+	+	+	+										
I									+	+	+	+	+	+	+	+	+	+							
J										+	–	+	+	+	+	+									
K											+	+	+	+											
L												+	+	+	+	+	+	+	+	+	+				
M													+	+	+	+	+	+	+	+					
N														+	+	+	+	+							
O															+	+	+								
P																+	+	+	+	+					
Q																	+	+	+	+	+	+	+	+	+
R																		+							
S																			+	+					
T																				+	+	+	+		
U																					+	+	+	+	+
V																									
W																							+		
X																								+	+

Table 8. Significant Spearman rank correlation coefficients among all plant species in the *Ecuador Right Sugarcane Plantation* over the first 10 years of succession. Significant positive associations are indicated by a "+" and significant negative associations are indicated by a "–". Plant species and families are indicated as follows: *Acalypha pladichephalus* (A), Asteraceae (B), *Digitaria sanguinalis* (C), Polypodiaceae (D), *Nectandra* spp. (E), *Stachys micheliana* (F), Piperaceae (G), *Lantana camara* (H), Verbenaceae (I), *Erythrina megistophyla* (J), *Piper aduncum* (K), *Rubus* spp. (L), *Commelina diffusa* (M), *Elephantopus mollis* (N), *Cecropia* spp. (O), *Costus* spp. (P), *Miconia* spp. (Q), Passifloraceae (R), Fabaceae (S), *Chusquea* spp. (T), Marantaceae (U), *Pilea* spp. (V), *Hieracium* spp. (W), *Sabicea* spp. (X), *Columnea* spp. (Y), and Orchidaceae (Z)

	B	C	D	E	F	G	H	I	J	K	L	M	N	O	P	Q	R	S	T	U	V	W	X	Y	Z
A	+	–	+	+	+	+	+	+	+	–	+	+	+	+											
B		+	+	+	+	+	+	+	–	+	+	+	+	+	+										
C			+	+	+	+	+	–	+	+	+														
D				+	–	–	+	–	+	+	+	+	+	+	+	+									
E					+	+	+	–	+	+	+														
F						+	+	+	+	+	+	+	+												
G							+	+	+	+															
H								+	+	+	+	+	+	+	+	+	+								
I									+	+	+	+	+	+											
J										+	+	+	+	+	+										
K											+	+	+	+	+										
L												–	+	+	+	+									
M													+	+	+	+	+	+							
N														+	+	+	+									
O															+	+									
P																+	+								
Q																	+	+	+	+					
R																		+	+	+	+				
S																			+	+	+	+	+	+	+
T																				+	+	+	+	+	
U																					+	+			
V																									
W																							+		
X																									

Table 9. Significant Spearman rank correlation coefficients among all plant species in the *Ecuador Left Sugarcane Plantation* over the first 10 years of succession. Significant positive associations are indicated by a "+" and significant negative associations are indicated by a "–". Plant species and families are indicated as follows: *Musa* spp. (A), *Costus* spp. (B), *Cuphea cartagenensis* (C), *Digitaria sanguinalis* (D), *Miconia* spp. (E), *Piper* spp. (F), *Rubus* spp. (G), *Sida rhombifolia* (H), Asteraceae (I), *Baccharis* spp. (J), Polypodiaceae (K), *Lantana camara* (L), *Vernonia patens* (M), *Acalypha pladichephalus* (N), *Solanum* spp. (O), *Saccharum officinarum* (P), *Piper aduncum* (Q), Verbenaceae (R), *Commelina diffusa* (S), *Erythrina megistophyla* (T), *Nectandra* spp. (U), *Altus* spp. (V), Orchidaceae (W), *Polybotria* spp. (X), Vernonia spp. (Y), and Polypodiaceae (Z)

	B	C	D	E	F	G	H	I	J	K	L	M	N	O	P	Q	R	S	T	U	V	W	X	Y	Z
A	+	+	–	+	–	+	+	+	+	+	+	+	+	–	+	+	+								
B		+	+	+	+	+	+	–	+	+	+	+	+	+	+	+									
C			+	+	+	+	+	+	+	+	+	+	+	+	+	+	+	+							
D				+	+	+	+	+	+	+	+	+													
E					+	+	+	–	+	–	+	+													
F						+	–	+	–	–	+	–	+	+											
G							+	+	+	+	+	+	+	+	+										
H								+	+	+	+	–	+	+											
I									+	+	+	+	+	+	+	+									
J										+	+	+	–	+											
K											+	+													
L												+	+	+	+	+	+	+	+	+					
M													+	+	+	+	+	+	+	+					
N														+	+	+	+	+	+	+	+				
O															+	+	+	+	+	+	+				
P																+	+	+	+						
Q																	+	+	+	+	+	+			
R																		+	+	+	+				
S																			+	+	+				
T																				+	+	+	+	+	+
U																					+				

Table 10. Significant Spearman rank correlation coefficients among all plant species in the *Ecuador Left Pasture* over the first 10 years of succession. Significant positive associations are indicated by a "+" and significant negative associations are indicated by a "–". Plant species and families are indicated as follows: *Setaria sphacelata* (A), *Acalypha pladichephalus* (B), *Xanthosoma sagittifolia* (C), *Heliocarpus americanus* (D), *Heliconia* spp. (E), *Ficus* spp. (F), Solanceae (G), Araceae (H), *Citrus* spp. (I), *Vernonia patens* (J), *Begonia* spp. (K), and *Nectandra* spp. (L)

	B	C	D	E	F	G	H	I	J	K	L
A	–	–	–	–	–	–	–	–	–	–	
B		+	+	+	+	+	+	+	+	+	+
C			+	+							
D											
E											
F						+	+	+			

Table 11. Significant Spearman rank correlation coefficients among all plant species in the *Ecuador Right Pasture* over the first 10 years of succession. Significant positive associations are indicated by a "+" and significant negative associations are indicated by a "–". Plant species and families are indicated as follows: *Setaria sphacelata* (A), *Erythrina megistophyla* (B), and Polpodiaceae (C)

	B	C
A	–	–

Further analysis of the plot data showed that the plots in Ecuador plantations show a significant log-normal pattern for the first 5 years with a flattening out after that. The plots in Ecuador pasture, however, show a significant geometric pattern over the first 7 years of succession. Pastures lag behind plantations and do not resemble them within this sampling time frame (Myster 2010). I also found that species richness had a significant positive relationship with productivity, where the slopes of the regression lines decreased over time for all fields taken together suggested the early upswing, leveling off, and later downswing of a unimodal curve; and this unimodal pattern held true both in the recovering sugarcane (*Saccharum officinarum* L.) and in the recovering Banana (*Musa* sp.) plantations. There was a delay, however, in the development of the unimodal pattern in the seeded pasture, perhaps due to root competition between the residual tussock grass (*Setaria sphacelata*) and Neotropical tree seedlings (Myster 2009).

Finally I used the plot data to investigate significant positive associations between pairs of old field species (Tables 6-11) were first computed and then clustered together into larger and larger species groups. I found that no pasture or plantation had more than 5% of the possible significant positive associations, clustering metrics showed groups of species participating in similar clusters among the five pasture/plantations over a gradient of decreasing association strength, and there was evidence for repeatable communities—especially after Banana cultivation—suggesting that past crops not only persist after abandonment but also form significant associations with invading plants (Myster 2012d).

Field experiments in these same six fields showed that most seeds were lost to predators, most of the remaining seeds were destroyed by seed pathogens, and most of those left germinated (Myster 2004a). There was more predation in the forest compared to the old field/forest border and more seed disease away from the forest compared to in it.

3. Case Study: Luquillo Experimental Forest (LEF), Puerto Rico USA

The Luquillo Experimental Forest of northeastern Puerto Rico, USA (18°20'N, 65°45'W) is one of a network of sites maintained by the long-term ecological research (LTER) program funded by the National Science Foundation. The LEF is found between 325 m and 1080 m in elevation (Myster and Fetcher 1990) and is classified as tropical lower montane wet forest below 600 m (Edmisten 1970, Walter 1979). It has rough, deeply dissected drainages with steep slopes. The average temperature is approximately 18 °C and the soils are derived from volcanoclastic primary substrate that has weathered into clay ultisols. The climate of LEF is dominated by trade winds that produce moisture-bearing clouds year round with a rainfall of between 2 m and 5 m per year. The wettest months are between July and October, and the driest months are between January and April. The dynamics of the LEF includes the natural disturbances of tree fall (Everham et al. 1996) and hurricanes but is also recovering from landslides (Myster and Walker 1995) and past agriculture (Myster 2012c).

3.1. LEF Lower Montane Wet Forest Floristics and Physical Structure

Within the LEF is the lower wet montane rainforest located up to an elevation of 600 m. Trees in this forest can exceed 30 m in height and have both a fairly continuous canopy and a moderately dense shaded understory. Those trees are most common on ridges and palms are common in riparian valleys. The most common tree species is *Dacryodes excelsa*. The other common tree species are *Casearia arborea*, *Manikara bidentata*, *Inga laurina* and *Sloanea berteriana*. These species and the palm *Prestoea montana* accounted for 65% of all stems at least 10 cm dbh in one sampling. The most common shrubs are *Palicourea riparia*, *Psychotria berteroana* and *Piper glabrescens*. (Brokaw et al. 2011). These tree species have relatively large seeds and are mainly obligately mycotrophic (Myster et al. 2013).

In two samplings of 1 ha plots in this forest, for all trees at least cm dbh, there were 434 and 936 stems per ha, 28.2 m^2 and 39.7 m^2 basal area per ha, 122 Mg and 300 Mg biomass per ha, and 33 and 45 tree species per ha. Total aboveground NPP of 10.5 Mg per ha per year (Brokaw et al. 2011). In LEF forest and gap field experiments I found that high light and litter seem to be universally good cues for regeneration, increasing seed/seedling survival for all species except for *Cecropia* whose small seeds may not be able to penetrate the litter layer. In addition, the high temperature environment in both artificially shaded and nonshaded areas of a natural gap exhibits less seed loss, an increase in the percent and rate of germination, and an increase in seedling survival for *Dacryodes*, than the lower temperature environment under an intact canopy. Low soil water is also a good cue for *Dacryodes* germination as it is for *Prestoea* and *Cecropia*. Finally, the lower temperature environment found under the forest canopy (compared to the natural gap) leads to less seed loss and more germination for *Guarea* (a mid-successional species). Results suggest that a good patch for regeneration of many species in this forest would have high light and a litter layer that moderates temperature and moisture extremes (Everham et al. 1996). Finally LEF studies suggest that if tree seeds could escape seed predation and seed pathogens, a high percentage would germinate (Myster and Everham 1999).

3.2. Regeneration of the LEF Lower Montane Wet Forest After Use as a Pasture and After Coffee Cultivation

It has long been known that the establishment and regular sampling of permanent vegetative plots, set up at the time of the abandonment, provides the best information about the nature of vegetative recovery and dynamics of fields by showing the actual sequences of plant abundance and species change over time (Pickett 1982). The patterns revealed from these plot data are important in their own right, but they also help to guide field experimentation (Myster 1993, Myster 2007a) and provide parameters for modeling (Myster 2010b). Perhaps most importantly we may observe, by careful mapping of individual plants within these fields (Myster 2003b), the plant-plant replacement process which creates all subsequent population and community patterns (Myster 2012b). In these recovering fields, replacements are working everywhere and at all times.

A long-term permanent plot study was set up in a natural, unseeded pasture in Puerto Rico that had been grazed for decades before it was abandoned. This study pasture borders the LEF close to the town of Sabana. Twenty-five 5 m×2 m contiguous plots (the total plot was 25m x 10m) were laid out on the border with the LEF in 1996 (Myster 2003b) with the long side parallel to the forest in order to maximize any forest edge effects. There were not any remnant trees or sprouting tree roots in the plots at the beginning of the study. Starting in May of 1997 and continuing annually for 10 years, the plots were sampled for percent cover of each plant species. Percent cover — an indication of a species' ability to capture light and, therefore, to dominate these areas that are in the process of becoming forested communities (Myster 2003b) — was estimated visually in relation to each plot's area.

One of the most important early issues is the nature and extent of any woody vegetation still remaining in the field after clearing from forest. Trees can sprout from coppiced tree trunks or root stocks, for example, and then function as "nurse" trees facilitating seed dispersal and perhaps other ways such as shading and killing lower growing forbs and grasses. Forest trees or shrubs may also spread into the field asexually (e.g., *Miconia* spp. in Puerto

Rican fields) from the woody remnants allowing trees to invade quickly from the forest edges, not dependent on the seed bank or the survival of newly dispersed seed. Because of the intensity of the mechanisms which remove tree seeds, these woody functions may be a significant part of facilitation in these fields.

The first analysis of this data used only the first five years of pasture sampling and showed that (1) grass dominated peaking at year three but still constituting 75% of the total plant cover after five years, (2) forbs also peaked at year three and then declined to 40% after five years, and (3) both ferns and woody species increased steadily to 45% and 15% total cover respectively after five years (Myster 2003b). During these first five years of succession, total percent cover ranged between 160% and 300% due to strata formation, percent species turnover declined to 25% in year four, species richness plateaued at 20 and species evenness stayed around 0.75. Focusing only on the trees and shrubs, the woody plants were present in the first sampling (Table 12) with greatest density close to the forest edge. The dominant species were the trees *Syzygium jambos* (an exotic species), *Calophyllum calaba*, and *Tabebuia heterophylla* and the dominant shrubs *Clidemia hirta* and *Miconia* spp., all peaking in abundant individualistically (Pickett 1982) in different years. Together these woody species showed (1) a maximum productivity of 400 $g/m^2/yr$ and a maximum total number of stems of 300, both in year two, (2) an increasing mean height of 240 cm in year five, and (3) a plateauing of total basal area of 950 cm^2 after year two (Myster 2003b). The spatial locations of each tree by basal diameter and species in year five were mapped and showed that trees invaded and grew from the forest edge, again in an individualistic manner.

The first 10 years of sampled pasture plot data were used to implicate plant interactions by generating Spearman coefficients of rank associations (Table 13) because controlled field experiments performed elsewhere have confirmed that species that interact significantly also form significant associations with each other (Tirado and Pugnaire 2005).

The pasture in Puerto Rico had a total of 307 significant pair-wise positive associations between species and 55 significant pair-wise negative associations over the 10 year sampling time period. Because most of the significant associations were positive, facilitation may be more common than competition in these stressful, early successional environments (Callaway and Walker 1997). Care must be taken, however, not to fall into the Nietzsche's "Cornaro" misinterpretation, because species that associate together may not be reacting to each other but to some other species or environment factor.

Table 12. All tree (T) and shrub (S) stems, at least 50 cm in height, sampled in the Sabana pasture permanent plot during the first five years of succession

Species	Year 1	Year 2	Year 3	Year 4	Year 5
Syzgium jambos (T)	18	80	76	70	65
Calophyllum calaba (T)	5	45	31	18	8
Clidemia hirta (S)	1	39	29	17	9
Tabebuia heterophylla (T)	4	11	10	12	5
Miconia impetiolaris (S)	4	24	22	21	20
Miconia prasina (S)	3	39	29	22	19
Citrus spp. (T)	2	3	3	3	3
Guarea guidonia (T)	1	4	3	5	2
Eugenia pseudopsidium (T)	2	7	6	8	7
Myrcia splendens (T)	0	2	3	1	1
Presotea montana (T)	0	1	2	2	2
Inga laurina (T)	2	5	4	5	3
Ocotea leucoxylon (T)	0	0	1	2	1
Andira inermis (T)	0	2	5	8	6
Inga vera (T)	2	9	11	12	10
Casearia sylvestris (T)	1	3	1	4	4
Miconia racemosa (S)	3	12	13	7	5
Psychotria berteriana (S)	0	2	1	2	2
Piper hisperdium (S)	2	23	10	3	3

Species that formed the most positive associations with other plants were the trees *Syzgium jambos, Guarea guidonia*, *Ocotea leucoxylon*, and *Prestoea montana*. Species that formed the most negative associations with other plants were the grass *Panicum* spp., the fern *Gleichenia bifida*, and the tree *Tabebula heterophylla*.

Also using the first 10 years of sampling data from the pasture, I developed allometric above-ground biomass equations for each of the ten most abundant tree and shrub species, which account for over 90% of the total stems and over 90% of the total basal area. Above-ground biomass was then computed from those equations (Table 14) and the most productive species were *Syzygium jambos*, *Miconia impetiolaris*, *Miconia prasina*, *Calophyllum calaba*, and *Clibadium umbella*. This productivity data was then plotted against turnover.

Table 13. Significant spearmen rank correlation coefficients among the top 25 most common plant species in the Sabana pasture over the first 10 years of succession. Significant positive associations are indicated by a "+" and significant negative associations are indicated by a "-". There can be a maximum of 10 significant associations between any two species, one for each sampling year. Plant species and families are indicated as follows: A. *Bromelia* spp., B. *Guarea guidonia*, C. *Ocotea leucoxylon*, D. *Citrus frutus*, E. *Syzgium jambos*, F. *Desmodium* spp., G. *Gleichenia bifida*, H. *Inga laurina*, I. *Citrus limon*, J. *Casearia sylvestris*, K. *Prestoea montana*, L. *Calophyllum calaba*, M. *Miconia presina*, N. *Eugenia pseudopsidium*, O. *Tabebuia heterophylla*, P. *Eugenia malaceensis*, Q. *Piper hisperdium*, R. *Andira inermis*, S. *Psychotria brachiate*, T. *Miconia racemosa*, U. *Psychotria berteriana*, V. *Xanthosoma* spp., W. *Clidemia hirta*, X. *Panicum* spp., Y. *Myrcia splendens* and Z. *Ocotea sintenisii*

	B	C	D	E	F	G	H	I	J	K	L	M	N	O	P	Q	R	S	T	U	V	W	X	Y	Z
A	+ -	+	-	+ +		+		+ +	-	+	+ + +			+ +			+		-					-	+ +-
B		+ + + + + +	+ +	+ + + + + + +	+ + -	+ + +			+ +	+ + + + +	+ +	+	-	+ + +				+	+	+ + +	+- --	+ +	+ +-	+ + + +	+ +
C			+	+ + +		+	+	+ +	+ + + +	+ +	+ + +	+	+ + +	+ +	+ + +							+ +	+ + + +	+ + + +	+ +

	B	C	D	E	F	G	H	I	J	K	L	M	N	O	P	Q	R	S	T	U	V	W	X	Y	Z
D				+	+ +	+ +			+	+	+		+	+ -		+ +	+		+						
E					+ -	+	+ + + +	+ +		+ + + + +	+ + + + + + +	+ +	+	+ + -	+ +	- - -	+	-	+ +	+ +	+ + +-	+-	+ + + +	+ + +	+ +
F						+ +		+		+	-				-		+					+	+-		
G							+	+ +	+ +	+	+ + + +	-		- -	+	+	- -	+ + +		+	--	+	+ +	--	+
H										+ + + + +		+	+ + +	- -	+ + + +			+		+	+	+		-	+ +
I										+ +	+	+		+ +	+	+ + +	+	+	+			+	+	+	+ + + + +
J											+ +		+ +	+ +	+				+ + + + +				+ +	+ +	+ +

Table 13. (Continued)

	B	C	D	E	F	G	H	I	J	K	L	M	N	O	P	Q	R	S	T	U	V	W	X	Y	Z
K														+ + -	+ +	+ +	+ -	+	+	+ +		+ + +	+-	+ +	+ + +
L												+ +	+	+	+ +	+			+ + +			-- -	+ +- -	+ + +	+
M														+ + + +	+ +		+	+ + + +	+			+ +	+		
N														+ +	+					+		+ +	+ +		
O																	- -	-	+		+-		-	-	+
P																			+			+ +	+	+ + +	+
Q																									
R																						+ +			
S																				+	-	+			

	B	C	D	E	F	G	H	I	J	K	L	M	N	O	P	Q	R	S	T	U	V	W	X	Y	Z
T																					+	+ + + +		+ +	+
U																									
V																							+ +-	+ +	
W																									+
X																								+ + + +	+
Y																									+

Table 14. Estimates of above-ground biomass (in g/m^2) for each of the ten most common tree species in each of the ten years of the study

Species	yr 1	yr 2	yr 3	yr 4	yr 5	yr 6	yr 7	yr 8	yr 9	yr 10
Syzygium jambos	35.1	202.4	77.7	110.5	102.2	89.9	82.2	76.6	85.7	77.9
Psychotria berteriana	0.8	0.7	1.7	0.7	2.7	3.2	2.9	3.3	4.5	5.8
Piper glabrescens	2.1	8.3	7.6	6.1	12.2	9.8	12.5	9.4	13.6	14.4
Miconia impetiolaris	17.2	37.6	21.8	31.1	25.8	26.6	29.3	33.3	37.6	44.8
Miconia prasina	13.3	53.4	25.6	33.9	26.6	28.4	27.7	36.9	42.2	48.9
Calophyllum calaba	9.5	115.5	36.9	20.6	21.5	20.5	19.9	15.6	22.3	18.8
Miconia racemosa	0.7	3.3	3.4	3.2	2.3	3.3	4.5	4.2	3.1	3.9
Clibadium umbella	21.1	49.4	26.1	38.5	37.9	44.4	48.8	52.2	46.2	54.3
Piper hispidum	1.1	7.3	7.1	6.2	13.7	14.4	14.8	15.7	15.1	16.8
Inga vera	2.2	24.4	2.8	4.1	7.4	9.8	12.2	13.8	15.5	17.7

The first two years of succession showed a slightly positive slope between productivity and turnover, but starting in year three there was a significant negative slope which became more negative with time. Turnover was also plotted against richness. There was a slightly negative slope between richness and turnover in the first year which became more negative with time (author, unpub. data).

The same plot design and sampling protocol was employed in a sun Coffee plantation located on private land in 1996.

The plots were sampled in 1997 and in 1998, but could not be continued because the farmer wanted the land back for other purposes. Plots showed that the Coffee trees themselves persisted, having 30% cover in year one and 23%

cover in year two. Like the pasture plot, grass dominated (94% in year one, 90% in year two) and *Ichnanthus pallens* most common, with forbs (45%, 50%: *Phytolacca icosandra* and *Musa* spp. most common) and woody vines (64%, 67%: same species as in the pasture plots) also common. Ferns were at low cover (5% in year one, 6% in year two: same species as in the pasture plots) but at greater levels than in the pasture plots (Myster 2003b). Trees were much less common after Coffee cultivation than in the pasture plots and the three most dominant trees and shrubs found in pasture were absent in Coffee. However, *Tabebuia heterophylla*, *Miconia* spp., and fruit trees were just as common. *Andira inermis* and *Guarea guidonia* attained greater cover in Coffee than in pasture but *Inga* spp. was not found. There were several species unique to the Coffee plots (Myster 2003b) and *Cecropia schreberiana* was again absent.

Again the same plot design and sampling protocol was employed in four pastures of different starting ages located close to the LEF: CoRd (5 years old), Los Davilas (13 years old), CoRd 9948 (18 years old) and Puente Roto (23 years old). These four pastures were first sampled in 2000 and then sampled annually.

The three permanent plot data sets (Sabana pasture, Coffee plantation, the four pastures that start in different years) were then used for an analysis of dominance-diversity curves. There was a significant log-normal pattern regardless if the plots were in a coffee plantation or in the Sabana pasture. The Sabana, CoRd990, CoRd9948, and Puente Roto pastures showed this pattern flattening over time with only the Los Davilas pasture maintaining a significant log-normal pattern (Myster 2010). Also, there was no lag time seen for any pasture compared to Coffee plantation.

Finally the first 10 years of sampling in the permanent plots of the Sabana pasture was combined with the first 5 years of sampling in these other four pastures, started at different times. This hybrid study, combining permanent plot data with chronosequences, hoped to provide enough overlapping of sampling years and pastures to predict succession in these pastures better than using chronosequences by themselves. Total plant richness and total plant cover did form long continuous patterns among the samplings, but species composition was too unique to each pasture to be able to combine into a single timeline. Consequently this approach may work best for broad structural parameters, but not for predicting individual species abundances (Myster and Malahy 2008).

In field experiments, tree seeds have been found to be rare in the soil seed bank of pastures and fields (Zimmerman et al. 2000, Cubina and Aide 2001,

Myster 2004a). They have been somewhat more common in the seed rain, but decrease sharply with distance from the forest edge in young pastures (Cubina and Aide 2001).

This is to be expected because early trees are more wind-dispersed than later-invading species and also any seed-dispersing animals avoid the open field, preferring areas next to forests and woody patches for perching and hiding, creating a positive feedback loop for tree dispersal which slows tree invasion out into the field. Woody tree density was only 0.01 seeds per m^2 after six months in a pasture (Zimmerman et al. 2000) and they were a small part of the appr. 20 plant seeds per m^2 sampled in young Puerto Rican fields after three months (Myster 2004a). They did increase, however, under shrubs growing naturally in the pasture (Myster 2006).

After tree seeds are dispersed, they face a host of tropical predators where even quick germination without dormancy may not save them. Detailed studies performed in Puerto Rico have always shown huge and immediate seed losses to predators including 50-75% lost after 14 days in Coffee plantation and pasture (Myster 2003a), 50-79% lost after 14 days in a pasture (Myster 2003c) and 67-90% lost after 14 days in various fields (Myster 2004a) . Within those general results, however, are important sources of variation (SOV) that show how some seeds survive. Those SOV include species-specific predator preferences, litter effects and different starting seed densities, field-types, patch-types seasons, ages, and seed sizes. In particular results imply that the density of predators — ants and small mammals have been implicated (Myster 2003c) — and their behavior and availability are affected by the structure of plant cover.

Data from the Puerto Rican Sabana pasture show SOV in these significant main effects in (1) species (*Syzgium jambos* had 50% losses, *Solanum tovum* had 65% losses, *Miconia prasina* had 76% losses, and *Piper aduncum* had 79% losses), (2) density (78% losses at 5 seeds per dish, 69% losses at 10 seeds per dish, 51% losses at 20 seeds per dish), and (3) patch type (grass patches had 69% losses, *Miconia* spp. patches had 65% losses). There were also SOV in these significant interaction effects (1) between species and density for *P. aduncum* and *S. tovum* (88% losses at low density vs. 85% losses at medium density vs. 56% losses at high density, 83% losses at low density vs. 70% losses at medium density vs. 27% losses at high density, respectively), and (2) between species and patch-type for *M. prasina* (93% losses in grass patches vs. 55% losses in *Miconia* spp. patches). Another experiment in the same pastures and in a Puerto Rican Coffee plantation again showed large species differences (*Guarea guidonia* had 34% losses,

Gonzalugunia spicata had 55% losses, *Miconia racemosa* had 66% losses, and *Casearia arborea* had 72% losses), with predators taking seeds at spatial scales of 2–11 cm (Myster 2003a).

Seeds lost to pathogenic diseases have always been much less than those lost to predation, in the ranges 20-45% after 14 days in Coffee plantation and Sabana pasture (Myster 2003a) and 5-10% 21 days in various fields (Myster 2004a) but may be more important in places such as sugarcane plantations with large mats of decaying cane biomass. Seed pathogens showed no clear spatial preferences in the Coffee plantation (Myster 2003a). Germination percentages were low for most species, 15-35% after 14 days in a Coffee plantation and pasture (Myster 2003a), 5-10% in various fields (Myster 2004a), and 2-90% in a pasture which was reduced by removal of vegetation (Zimmerman et al. 2000). Germination occurred at larger spatial scales than predation, 11-23cm, in the Coffee plantation (Myster 2003a). Greenhouse and shadehouse studies, however, suggest that these percentages could be greater in other fields and studies (Myster and Everham 1999, Zimmerman et al. 2000) and that many tree seeds could germinate if they can escape predation and pathogens.

Detailed studies performed in Puerto Rico have shown 30-50% survival of tree seedlings after 6 months in various fields (Myster 2004a) and 2-90% survival after 6 months in a pasture (Zimmerman et al. 2000) with *Guarea guidonia*, *Calophyllum calaba*, and *Andira inermis* surviving best, *Psychotria brachiata* least. There was also significant SOV among pasture tree species in (1) survival: *Piper aduncum* 21%, *Cecropia schreberiana* 34%, *Guarea guidonia* 45%, *Veronica patens* 47%, *Solanum torvum* 51% and *Miconia prasina* 55% and (2) growth parameters: *Cecropia* (85cm in mean height, 1.18 cm mean basal diameter, 3.1 leaves on average, 350 cm^2 mean leaf area), *Solanum* (90 cm, 0.76 cm, 5.5, 150 cm^2), *Piper* (66 cm, 0.82 cm, 5.9, 75 cm2), *Veronica (*95 cm,0.92, 7.2, 104 cm2) and *Miconia* ((72 cm, 0.75, 4.9,122 cm2). In general, seedlings had fatter stems and more leaves in the forest/pasture border and taller stems 5-10 m from the forest edge. Competition for light is implicated as a major mechanism affecting seedling establishment in fields and pastures, because forest trees compete primarily for light, but competition with grasses for water and soil nutrients may be more important in early succession. Herbivory levels (< 2% of leaf area on average) and pathogens levels (even less: Myster 2004a) are perhaps too low even to affect competition (Myster 1993).

4. Summary and Suggestions for Future Research

More sampling and field experiments need to be done both at Maquipucuna Reserve and at the LEF, including permanent plots and seed/seedlings field experiments.

The work already done, however, allows for some comparison between primary forests and secondary forests recovering from agriculture in these three areas: (1) floristics, (2) physical structure, and (3) regeneration after treefall in gaps vs. after agriculture.

1. Floristics: The fields at Maquipucuna are 17 years since abandonment. As such they do have a few of the primary cloud forest species in common (*Cecropia* sp., *Piper* sp.), however these species are not yet dominant and inhibitory effects of the past crop on tree invasion and growth (Myster 2012c) may linger on for a few more years. The Sabana pasture at LEF is 10 years old at last sampling but does contain *Presotea montana* and *Inga laurina* both dominant trees in the LEF. The dominant LEF shrub *Psychotria berteriana* is also present in the pasture. Tree species found after tree fall (Devoe 1989) and landslide (Myster and Walker 1997) in the LEF, and thus thought to be common "pioneer" trees (e.g., *Cecropia* spp., *Shefflera* spp.), however, were not present in the Sabana pasture data.
2. Physical structure: The cloud forest at Maquipucuna (1 ha plot sampled for trees down to 10 cm dbh) had 294 stems, a biodiversity of 25 species and 11.2 m^2 basal area. The Maquipucuna old fields (0.004 ha plot sampled for trees down to 1 cm dbh) had 200 stems, 20 species and a basal area 0.10 m^2. In two samplings of 1 ha plots in the LEF (for all trees at least 10 cm dbh) there were 434 and 936 stems per ha, 28.2 m^2 and 39.7 m^2 basal area per ha, and 33 and 45 tree species per ha (Brokaw et al. 2011). Production in the LEF is 11-14 $kg/m^2/yr$ and 12.2 $kg/m^2/yr$ one year after Hurricane Hugo but pasture productivity may take 55 years to recover to forest levels (Aide et al. 1995).
3. Regeneration: Comparison is not possible between Maquipucuna and its old fields because work has yet to be done in the Cloud forest (scheduled for years 2014, 2016, 2018). But at LEF we can compare two species found both in the Sabana pasture and in LEF field studies (*Cecropia* sp., *Guarea* sp.). In the LEF *Cecropia* seeds do well in high light and low soil water, but not under thick litter. Also, *Guarea* seeds

survive and germinate better under a forest canopy. In the LEF Coffee plantation *Guarea* lost 34% of its seeds to seed predators, *Cecropia* seedings had 34% survivorship with *Guarea at* 45%. In addition *Cecropia* sp.seedlings grew well at 85cm in mean height, 1.18 cm mean basal diameter, 3.1 leaves on average and 350 cm^2 mean leaf area.

Restoration and management is necessary to return abandoned crop land to high productivity in as short a period of time as possible and then maintain its sustainability. One way to manage succession towards an end that better satisfies human needs is to retard succession. Another way is to quicken the rate of recovery, perhaps by increasing facilitating mechanisms, such as dispersal, and decreasing inhibitory mechanisms, such as predation and competition (as discussed above), in order to generate rainforest timber to be used as a sustainable resource or to create rainforest habitat for rare species. One more option is to hold succession in a state when agriculture is less costly to maintain. Yet another approach is to alter the natural successional trajectory itself to match soil conditions for maximum crop production. Mechanisms and tolerances, as discussed, can be manipulated either as inhibitors or facilitators towards any of these ends.

Whatever restoration approach is used, studies of post-agricultural succession are indispensable to the success of restoration attempts because they provide information about how Neotropic areas regenerate and function. Management of old field succession has to have scientific underpinnings in order to be effective.

Hence, the successional processes and mechanisms that have been discussed in this chapter can serve any restoration effort.

All the data from permanent plot studies related in this chapter are archived at the Luquillo Experimental Forest as LTERDATB #97, LTERDATB #100, LTERDATB#101 and LTERDBAS#109, as part of the long term ecological research (LTER) program funded by the US National Science Foundation. One may visit their website (http://luq.lternet.edu) for further details.

The LTER program has been a continuous part of the University of Puerto Rico ad U.S. Forest Service in Puerto Rico since 1988.

Acknowledgments

I thank Arcenio Barrenas (Maquipucuna) and Maria Aponte (LEF) for their help in the field. I also thank Eda Meléndez and her staff for managing the LTER data sets. All research was performed under grants BSR-8811902 and DEB-9411973 from the National Science Foundation to the Institute for Tropical Ecosystem Studies, University of Puerto Rico, and to the USDA Forest Service International Institute of Tropical Forestry as part of the Long-Term Ecological Research Program in the Luquillo Experimental Forest. Additional support was provided by the Forest Service and the University of Puerto Rico.

References

Aide, T. M., and Cavelier, J. 1994. Barriers to lowland tropical forest restoration in the Sierra Nevada de Santa Marta, Colombia. *Restoration Ecology,* 2:219–229.

Bazzaz, F. A. 1968. Succession on abandoned fields in the Shawnee Hills, southern Illinois. , 49(5):924–936.

Bazzaz, F. A. 1996. Plants in Changing Environments: Linking Physiological, Population, and Community Ecology. Cambridge: Cambridge University Press.

Borhidi, A. 1988. Vegetation dynamics of the savannization process on Cuba. *Vegetatio,* 77:177–183.

Brokaw, N. V. L. 1982. The definition of treefall gap and its effect on measures of forest dynamics. *Biotropica,* 14:158–160.

Brokaw, N., Crowl, T. A., Lugo, A. E., McDowell, W. H., Scatena, F. N., Waide, R. B., and M. R. Willig (eds) 2012. A Caribbean Forest Tapestry: the multidimentional nature of disturbance and response. Oxford University Press, Oxford, UK.

Brown, S. and Lugo, A. E. 1990b. Effects of forest clearing and succession on the carbon and nitrogen content of soils in Puerto Rico and US Virgin islands. *Plant and soil,* 124:53-64.

Buchholz, T., T. Tennigkeit and Weinreich, A. 2004. Maesopsis eminii – a challenging timber tree species in Uganda – a production model for commercial forestry and smallholders. Proceedings of the international union of forestry research organizations (IUFRO) conference on the economics and management of high productivity plantations, Lugo, Spain, 15p.

Buell, M. F., Buell, H. F., Small, J. A., and Siccama, T. G. 1971. Invasion of trees in secondary succession on the New Jersey piedmont. *Bulletin of the Torrey Botanical Club,* 98:67–74.

Busing, R. T. 1996. Estimation of tree replacement patterns in an Appalachian *Picea-Abies* forest. *Journal of Vegetation Science,* 7:685–694.

Buschbacher, R. J. 1986. Tropical deforestation and pasture development. *BioScience,* 36:2 –28.

Callaway, R. M. and Walker, L. R 1997. Competition and facilitation: a synthetic approach to interactions in plant communities. *Ecology,* 78:1958-1965.

Castro, I. A., Sterling, S., and Galiano, E. F. 1986. Multi-species pattern analysis of Mediterranean pastures in three stages of ecological succession. *Vegetatio,* 68:37–42.

Cubina, A. and Aide, T. M. 2001. The effect of distance from forest edge on seed rain and soil seed bank in a tropical pasture. *Biotropica,* 33: 260-267.

Devoe, N. N. 1989. Differential seedling and regeneration in openings and beneath closed canopy in sub-tropical wet forest. Ph. D. dissertation. Yale University, New Haven, CO.

Donfack, P., Floret, C., and Pontanier, R. 1995. Secondary succession in abandoned fields of dry tropical northern Cameroon. *Journal of Vegetation Science,* 6:499–508.

Edmisten, J. 1970. Some autoecological studies of Ormosia krugii. In A Tropical Rain Forest, ed. H. T. Odum and R. F. Pigeon, Chapter D-8. Springfield, Virginia: National Technical Information Service.

Egler, F. E. 1954. Vegetation science concepts. I. Initial floristic composition, a factor in oldfield vegetation development. *Vegetatio,* 4:412–417.

Everham, E. M. III, R. W. Myster and E. VanDeGenachte. 1996. Effects of light, moisture, temperature, and litter on the regeneration of five tree species in the tropical montane wet forest of Puerto Rico. *American Journal of Botany,* 83:1063-1068.

Fearnside, P. M. S. 1993. Deforestation in Brazilian Amazonia: the effect of population and land tenure. *Ambio*, 22:537–545.

Fernandes, D. N., and Sanford Jr., R. F. 1995. Effects of recent land-use practices on soil nutrients and succession under tropical wet forest in Costa Rica. *Conservation Biology,* 9:915 –922.

Finegan, B. 1984. Forest succession. *Nature,* 312:109–114.

Gentry, A. 1993. A field guide to woody plants of northwest South America (Colombia, Ecuador, Peru). Conservation International, Washington, DC

Grainger, A. 1988. Estimating areas of degraded tropical lands requiring replenishment of forest cover. *International Tree Crops Journal,* 5:31–61.

Grubb, P. J. 1977. The maintenance of species richness in plant communities: the importance of the regeneration niche. *Biological review of the Cambridge Philosophical Society,* 52:107 –145.

Grubb, P. J., Lloyd, J. R., Pennington, T. D. A. and Whitmore, T. C. 1963. A comparison of montane and lowland rain forest in Ecuador. I. The forest structure, physiognomy, and floristics. *J. Ecol*., 51:567-601.

Harper, J. L. 1977. Population Biology of Plants. London: Academic Press.

Heaney, A. and J. Protor. 1990. Preliminary studies on forest structure and flostistics on Volcan Barva, Costa Rica. *J. Trop. Ecol*., 6:307-320.

Hobbie, S. E. 1992. Effects of plant species on nutrient cycling. *Trends in Ecology and Evolution,* 7:336–339.

Keller, M., Veldkamp, E. Weitz, A. M., and Reiners, W. A. 1993. Effect of pasture age on soil trace-gas emissions from a deforested area of Costa Rica. *Nature,* 365:244–246.

Ludwig, JA and J. F. Reynolds 1988. Statistical ecology. John Wiley & sons, NY.

Lugo, A. E. 1992. Comparison of tropical tree plantations with secondary forests of similar age. *Ecological Monographs,* 62:1–41.

Mabberley, D. J. 1992. Tropical Rain Forest Ecology. New York: Chapman and Hall.

Miles, J. 1979. Vegetation Dynamics. New York: John Wiley Myers, N. 1991. Tropical deforestation: the last situation. *BioScience,* 41:282.

Myster, R. W. 1993. Tree invasion and establishment in old fields at Hutcheson Memorial Forest. *The Botanical Review,* 59:251–272.

Myster, R. W. 2003a. Seed regeneration mechanisms over fine spatial scales on recovering Coffee plantation and pasture in Puerto Rico. *Plant Ecology,* 166(2):199–205.

Myster, R. W. 2003b. Vegetation dynamics of a permanent pasture plot in Puerto Rico. *Biotropica,* 35:422–428.

Myster, R. W. 2003c. Effects of species, density, patch-type and season on post-dispersal seed predation in a Puerto Rican pasture. *Biotropica,* 35:542–546.

Myster, R. W. 2004a. Regeneration filters in post-agricultural fields of Puerto Rico and Ecuador. *Plant Ecology,* 172:199–209.

Myster, R. W. 2004b. Post-agricultural invasion, establishment and growth of Neotropical trees. *The Botanical Review,* 70:381–402.
Myster, R. W. 2006. Shrub vs. grass patch effects on the seed rain and seed bank of a five-year pasture in Puerto Rico. *Ecotropica,* 12:5-12.
Myster, R. W. 2007a. Early successional pattern and process after Sugarcane, Banana and pasture cultivation in Ecuador. *The New Zealand Journal of Botany,* 46:101-110.
Myster, R. W. 2007b. Introduction. Pp 3-21 n *Post-agricultural succession in the Neotropics*. Myster, R W. (ed). Springer-Verlag, Berlin.
Myster, R. W. 2009. Are productivity and richness consistently related after different crops in the Neotropics? *Botany,* 87:357-362
Myster, R. W. 2010a. Testing dominance-diversity hypotheses using data from abandoned plantations and pastures in Puerto Rico and Ecuador. *Journal of Tropical Ecology,* 26:247 -250.
Myster, R. W. 2010b. A comparison of tree replacement models in oldfields at Hutchenson Memorial Forest. *The Journal of the Torrey Botanical Society,* 137:113-119.
Myster, R. W. 2012a. Long-term data from fields recovering after Sugarcane, Banana and Pasture cultivation in Ecuador. Dataset papers in Ecology Volume 2013, Article ID 468973.10 pages, DOI 10.7167/2013/46873.
Myster, RW. 2012b. Plants replacing plants: the future of community modeling and research. *The Botanical Review,* 78:2-9.
Myster, R. W. 2012c. Post-agricultural ecotones in Puerto Rico. Pp 111-137 in Ecotones between forest and grassland. Myster, R. W. (ed) Springer-Verlag, Berlin.
Myster, R. W. 2012d. A refined methodology for defining plant communities using data after agriculture in the Neotropics. *The Scientific World Journal.*
Myster, R. W. and Everham III, E. M. 1999. Germination cues across the disturbance regime in the Puerto Rican rainforest. *Tropical Ecology,* 40:89-98.
Myster, R. W. and N. Fetcher. 2005. Ecotypic differentiation and plant growth in the Luquillo Mountains of Puerto Rico. *Journal of Tropical Forest Science,* 17:163-169.
Myster, R. W., L. Lebron, A. B. Portugal Loayza and J. K. Zimmerman. 2013. Mycotrophic strategy of 13 common Neotropical trees and shrubs. *Journal of Tropical Forest Science,* 25:34-41.

Myster, R. W. and M. P. Malahy. 2008. Is there a middle-way between permanent plots and chronosequences? *Canadian Journal of Forest Research,* 38:3133-3138

Myster, R. W., and McCarthy, B. C. 1989. Effects of herbivory and competition on survival of Carya tomentosa (Juglandaceae) seedlings. *Oikos,* 56:145–148.

Myster, R. W., and Pickett, S. T. A. 1988. Individualistic patterns of annuals and biennials in early successional old fields. *Vegetatio,* 78:53–60.

Myster, R. W., and Pickett, S. T. A. 1990a. The Buell successional study at Hutcheson Memorial Forest Center. Permanent Plotter 3:2–3.

Myster, R. W., and Pickett, S. T. A. 1990b. Initial conditions, history and successional pathways in ten contrasting old fields. *American Midland Naturalist,* 124:213–238.

Myster, RW and S. T. A. Pickett. 1992. Effects of palatability and dispersal mode on spatial patterns of tree seedlings in old fields. *Bulletin of the Torrey Botanical Club,* 119:145-151.

Myster, R. W., and Pickett, S. T. A. 1993. Effects of litter, distance, density and vegetation patch type on postdispersal tree seed predation in old fields. *Oikos,* 66:381–388.

Myster, R. W., and Pickett, S. T. A. 1994. A comparison of rate of succession over 18 yr in ten contrasting old fields. *Ecology,* 75:387–392.

Myster, R. W., and Sarmiento, F. O. 1998. Seed inputs to microsite patch recovery on two Tropandean landslides in Ecuador. *Restoration Ecology,* 6:35–43.

Myster, R. W. and L. R. Walker. 1997. Plant successional pathways on Puerto Rican landslides. *Journal of Tropical Ecology,* 13:165-173.

Nadkarni, N. M., Matelson, T. J. and W. A. Haber. 1995. Structural characteristics and floristic composition of a Neotropical cloud forest, Monteverde, Costa Rica. *J. Trop. Ecol.,* 11:481-495.

Nascimento, HEM and W. F. Laurance. 2001. Total aboveground biomass in central Amazonian rainforests: a landscape-scale study. *Forest Ecology and Management,* 157:1-11.

Oosting, H. J. 1942. An ecological analysis of the plant communities of Piedmont, North Carolina. *American Midland Naturalist,* 28:1–126.

Padoch, C., and Vayda, A. P. 1983. Patterns of resource use and human settlement in tropical forest. In Tropical Rain Forest Ecosystems. Volume A, ed. F. B. Golley, pp. 301–313. New York: Elsevier.

Pickett, S. T. A. 1982. Population patterns through twenty years of old field succession. *Vegetatio,* 49:45–59.

Pickett, S. T. A., and White, P. S. 1985. The Ecology of Natural Disturbance and Patch Dynamics. Orlando, FL: Academic Press.

Raunkaier, C. 1934. The Life forms of Plants and Statistical Plant Geography . Oxford, UK: Clarendon.

Rhoades, C. C., Eckert, G. E., and D. C. Coleman. 1998. Effect of pasture trees on soil nitrogen and organic matter: implications for tropical montane forest restoration. *Restoration Ecology,* 6:262–270.

Rhoades, C. C., and Coleman, D. C. 1999. Nitrogen mineralization and nitrification following land conversion in montane Ecuador. *Soil Biology and Biochemistry,* 31: 1347 –1354.

Romoleroux, K, Foster, R, Valencia, R, Condit, R, Balslev, H and E. Losos. 1997. Especies lenosas (dap => 1 cm) encontradas en dos hectareas de un bosque de la Amazonia ecuatoriana. Pp 189-215 in Valencia, R. and H. Balslev (eds) Estudios Sobre Diversidad y

Ecologia de Plantas. Pontificia Universidad Catolica del Ecuador, Quito, Ecuador.Sarmiento, F.O. 1995. Naming and knowing a Tropandean landscape: The Maquipucuna Reserve as a case study. *The George Wright Forum,* 12: 15–22.

Sarmiento, F. O. 1997. Arrested succession in pastures hinders regeneration of Tropandean forests and shreds mountain landscapes. *Environmental Conservation,* 24:14–23.

Schwarzkopf, T., Riha, S. J., Fahey, T. J. and S. Degloria. 2011. Are cloud forest tree structure and environment related in the Venezuelan Andes? *Austral ecology,* 36:280-289.

Schultz, J. 2005. The Ecozones of the World, Springer, Berlin Heidelberg New York.

Serrao, E. S. A., and Toledo, J. M. 1990. The search for sustainability in Amazonian pastures. In Alternatives to Deforestation: Steps Toward Sustainable Use of the Amazon Rain Forest, ed. A. B. Anderson, pp. 195–214. New York: Columbia University Press.

Skole, D., and Tucker, C. 1993. Tropical deforestation and habitat fragmentation in the Amazon: Satellite data from 1978 to 1988. *Science,* 260:1905–1910.

Singh, K. D. 1993. The 1990 tropical forest resources assessment. Unasylva 174:10–19.

Staver, C. 1990. Why farmers rotate fields in maize-cassava-plantain bush fallow agriculture in the wet Peruivan Amazon. *Human Ecology,* 17:401–426.

Tanner, E. V. J. 1977. Four montane rain forests of Jamaica: a quantitative characterization of the floristics, the soils and the foliar mineral levels, and a discussion of the interrelations. *J. Ecol.*, 65:883-918.

Thomlinson, J. R., Serrano, M. I., del M. Lopez, T., Aide, T. M., and Zimmerman, J. K. 1996. Land-use dynamics in a post-agricultural Puerto Rican landscape (1936–1988). *Biotropica,* 28:525–536.

Tilman, D. 1988. Plant Strategies and the Dynamics and Structure of Plant Communities. Princeton, NJ: Princeton University Press.

Tirado, R., and Pugnaire, F. I. 2005. Community structure and positive interactions in constraining environments. *Oikos,* 111:437–444.

Uhl, C., Buschbacher, R., and Serrao, E. A. S. 1988. Abandoned pastures in eastern Amazonia. I. Patterns of plant succession. *Journal of Ecology,* 76:663–681.

Uhl, C. 1987. Factors controlling succession following slash-and-burn agriculture in Amazonia. *Journal of Ecology,* 75:377–407.

Vitousek, P. M 1984. Litterfall, Nutrient Cycling, and Nutrient Limitation in Tropical Forests. *Ecology,* 65:285–298

Walter, H. 1979. Vegetation of the Earth and the Ecological Systems of the Geo-biosphere. New York: Springer-Verlag.

Weaver, P. L., Medina, E., Pool, D., Dugger, K., Gonzales-Liboy, J. and E. Cuevas. 1986. Ecological observations in the dwarf Cloud forest of the Luquillo mountains in Puerto Rico. *Biotropica,* 18: 79-85.

Whittaker, R. H. 1975. Communities and Ecosystems. New York: Macmillan.

Zahawi, R. A., and Augspurger, C. K. 1999. Early plant succession in abandoned pastures in Ecuador. *Biotropica,* 31:123–129.

Zimmerman, J. K., J. B. Pascarella and Aide, T. W. 2000. Barriers to forest regeneration in an abandoned pasture in Puerto Rico. *Restoration ecology,* 8:350-360.

In: Forest Ecosystems
Editor: Noel C. Roberts
ISBN: 978-1-63117-815-3

Chapter 2

Modeling of Vegetation Dynamics and Spatial Pattern in the Declined Climax Mountain Norway Spruce Forest

Lubos Matejicek* and Marek Biskupic
Institute for Environmental Studies, Charles University in Prague, Faculty of Natural Science, Prague, Czech Republic

Abstract

Research dealing with modeling of vegetation dynamics and spatial pattern in a declined mountain spruce forest is closely related to remote sensing, which can discover changes in land cover. Assessing land cover by classification of archived images obtained from a series of Landsat sensors can provide information about the succession of a spruce forest. New approaches can be implemented on the basis of remote sensing, geographic information systems (GISs) and dynamic modeling. This offers a wide range of spatial and temporal analyses to explore spruce forest succession. The described research shows remote sensing data from Landsat sensors including: Thematic Mapper (TM), Enhanced Thematic Mapper Plus (ETM+) and Operational Land Imager (OLI) of Landsat 5,

---

* Corresponding author: Lubos Matejicek. Institute for Environmental Studies, Charles University in Prague, Faculty of Natural Science, Benatska 2, 128 01, Prague, Czech Republic. E-mail: lmatejic@natur.cuni.cz.

Landsat 7 and Landsat 8, respectively. The area of interest is located in the Tatra National Park, which is part of the Carpathian Mountains in Europe. The explored period from July 1994 to June 2010 is based on available archive images from Landsat 5 TM. An extension of the period to September 2013 can be obtained from archive images from Landsat 7 ETM+ and Landsat 8 OLI. The most important period was the winter season in November 2004 when the territory was swept by gale-force winds with gusts of almost 200 km/hour. In order to explore land cover changes for modelling vegetation dynamics and spatial patterns, unsupervised classification and supervised classification were used to detect land cover classes from image compositions based on Landsat 5 TM bands for the images captured in July 1994 and in June 2010. The target data structures for modelling vegetation dynamics and spatial pattern are formed by a grid of cells that represent each local site. The numerical solution of a spatio-temporal model for dominant seedlings of *Picea abies* is based on a set of differential equations for the grid of cells. The growth predictions of individual trees are based on a set of algebraic equations. The proposal for a spatio-temporal model in GIS enables other processing to provide prediction of spruce forest regeneration and succession after the winds in November 2004. GIS can integrate all the spatial data together with model predictions and support the decision-making processes in risk assessment analysis and natural conservation.

Keywords: Vegetation dynamics; spruce forest; remote sensing; GIS

1. Introduction

Improved predictions of vegetation dynamics and spatial patterns are required for a broad range of ecological applications that include studies of the terrestrial carbon cycle, research on biodiversity and wildlife habitat together with forest ecosystem services that are essential for human well-being, including forest timber production, watershed protection and recreational facilities. On a global scale, the role of mountain forests is linked to climate regulation, carbon sequestration, and surface radiation modulation. However, prediction of forest conditions depends on forest type and composition as well as on management and nature disturbances, making it paramount to monitor forests repeatedly and consistently with high spatial detail across an area of interest. In addition to a number of techniques in the field, remote sensing has great potential to provide more complex information on forest cover and its changes.

It can be used to model of forest vegetation dynamics and spatial patterns, and often the only source of information, because data on forest cover are not always publicly accessible.

Modelling of vegetation dynamics and spatial patterns constitutes an exacting task that involves processing of the input data, creating mathematical models, solving numerical models, validation and verification of model results, and application of model predictions. A number of models have been developed for short-term and long-term predictions on various spatial scales.

Gap models are widely used to predict forest succession [1, 2], but lack of appropriate empirical data renders model validation a non-trivial task. Species composition is often based on an equilibrium state. Thus, only the composition of simulated equilibrium species can be evaluated with limited dynamics.

Many other models deal with forest dynamics in the framework of local time series of stand development, which results in limited usability for other forest sites and the simulation period cannot exceed a few decades [3-5]. The described modelling of vegetation dynamics and spatial patterns in a mountain environment is focused on assessment of land cover changes in a period of a few decades. The landcover datasets are derived from archived Landsat images. The selected area of interest in the Tatra National Park in Europe is characterized by temporally changing landcover structures caused by human-induced deforestation and the bora windstorm in 2004. The time period of modelling for a selected site is given by the available series of Landsat images.

2. Remote Sensing Data and Procesing Methods

Remote sensing systems, particularly those deployed on satellites, provide a repetitive view of the earth's surface that can be invaluable for monitoring short-term and long-term changes. For passive remote sensing, the optical regime from visible through thermal depends on two sources of radiation: the sun and the earth. The radiation collected in the visible to shortwave infrared originates from the sun. Part of this radiation is reflected at the Earth's surface and part is scattered by the atmosphere, without ever reaching the Earth. The radiation collected in the thermal infrared region is emitted directly by materials on the Earth's surface and combines with self-emitted thermal radiation in the atmosphere as it propagates upward [6-8]. A wide range of applications in remote sensing has been carried out in the context of research projects focused on climate change, landscape dynamics, hazards and disasters [9-12].

This began in a wider scale when the Landsat Multispectral Scanner System (MSS) provided a consistent set of synoptic, high resolution earth images to the global scientific community. This modern era of remote sensing of the Earth by Landsat has been utilized since 1972 [13-16]. Many countries and multinational agencies operate remote sensing with various sensor parameters. An increasing number of new sensor systems present a continuing challenge for advanced data storage and image processing.

Since the late 1990s, commercial development of high-performance orbital sensors with resolution under one meter in the panchromatic mode has created new commercial markets and public service opportunities for satellite applications, such as real estate marketing, urban and transport planning, natural and man-made disaster mapping, agricultural management or environmental monitoring.

Some remote sensing system are restricted to use military intelligence [17-22]. The long-term remote sensing services are transformed into programs that are focused on three major components: remote sensing missions and data acquisition, long-term data preservation and access, and remote sensing research and data utilization. For example, the Landsat Program, a joint effort of USGS and NASA, was established to routinely gather land images from space. This program has resulted in a long-term record of natural and man-made changes in the global landscape. Across four decades since 1972, Landsat satellites have continuously acquired space-based images and have provided worldwide science and resource-management communities with an archive of space-based land remotely sensed data [23].

Nowadays, Landsat 5 (launched in 1984), Landsat 7 (launched in 1999) and Landsat 8 (launched in 2013) continue to capture hundreds of images of the Earth's surface each day. The improved Thematic Mapper (TM) sensors on board Landsats 4 and 5 is designed with several additional bands in the short wave infrared (SWIR) part of the spectrum, improved spatial resolution of 30 meters for the visible, near infrared (NIR) and SWIR bands, and the addition of a 120 meter (thermal radiation) TIR band. Landsat 7 carries the Enhanced Thematic Mapper Plus (ETM+), with 30-meter visible, NIR and SWIR bands, a 60-meter spatial-resolution TIR band and a 15-meter panchromatic band. The Landsat Data Continuity Mission (LDCM) is the next-generation Landsat satellite (Landsat 8) and ensures the continued acquisition and availability of Landsat-like data well beyond the duration of the current Landsat 5 and Landsat 7 missions. The Operational Land Imager (OLI) instrument advances future measurement capabilities while ensuring compatibility with historical data.

The OLI collects data in the visible, near infrared and shortwave infrared wavelength regions, as well as in a panchromatic band. Two new spectral bands have been added: a deep-blue band for coastal waters and aerosol studies, and a band for cirrus cloud detection. A Quality Assurance band is also included to indicate the presence of terrain shadowing, data artefacts and clouds. The Thermal Infrared Sensor (TIRS) was added to the payload to collect data in two long wavelength thermal infrared bands and has a 3-year design lifetime. The 100-meter spatial resolution of TIRS data is registered to the OLI data to create radiometrically and geometrically calibrated, terrain-corrected 16-bit Level 1 data products. NASA managed the satellite during the on-orbit initialization and verification period after launch. Once the post-launch assessment review and the mission transition reviews were completed, the USGS became responsible for the operations of the mission, including: health and safety of the spacecraft, orbital maintenance and also collecting, archiving, processing, and distributing data products. Landsat data held in the USGS archives are available for downloading free of charge and with no user restrictions via EarthExplorer http://earthexplorer.usgs.gov, GloVis http:// glovis.usgs.gov, or the LandsatLook Viewer at http://landsatlook.usgs.gov.

Detailed descriptions of Landsat operations, technology, and products are accessible via http://landsat.usgs.gov [24-26].

The first steps in processing of remote sensing data are mainly focused on pre-processing operations, such as feature extraction, radiometric corrections and geometric transformations. There is no definitive list of pre-processing techniques because each project requires specific attention and individual pre-processing decisions. The radiometric corrections adjust digital values for the effects of a hazy atmosphere. The geometric transformations are used to provide spatial correction by a digital terrain model and to bring an image into registration with selected map coordinates. Because pre-processing changes data, it is recommended to use only those operations that are essential to obtain specific results.

The next step in processing methods often includes image classification based on conventional classifiers for assigning pixels to classes. By comparing pixels to one another and to pixels of known identity, the classifier can assemble groups of similar pixels into classes that are associated with categories such as urban zones, industrial sites, agricultural land, forest, watersheds and water bodies. These classes create regions on an image which are transformed into a mosaic of uniform parcels, each identified by a colour. The pixels in each class are spectrally more similar to one another than they are to pixels in other classes.

For the mountain forest, a few subclasses can be identified as coniferous, mixed and broad-leaved forests together with grassland and water bodies. Classes can also be formed by sites with various conditions caused by human management and nature disturbances.

In addition to techniques for clustering pixels using various indexes, unsupervised classification and supervised classification represent the dominant groups of methods for image classification that deal with identification of natural groups within multispectral data.

Unsupervised classification aggregates each pixel into one of a specified number of clusters. Each cluster is statistically separate from the other clusters. The multivariate statistics establishing the cluster definition are stored in a signature file. After the unsupervised classification, each cluster has to be attached to a suitable land cover category, such as damaged and undamaged forest, grassland, water bodies etc. Supervised classification is based on sampling of the features, such as a coniferous forest, grassland, water bodies etc. Identification is made by enclosing the samples in a map with polygons. Once the training samples are established, multivariate statistics are calculated on this basis to establish the relationships within and between the classes. Like in the unsupervised classification, the statistics are saved in a signature file [27-30]. In case of spatio-temporal modelling in a mountain forest, the data inputs are based on Landsat images. The steps in preforming classifications include pre-processing of the input data with radiometric correction and a geometric transformation. In order to assess a long period of the mountain forest land cover changes in comparable vegetation seasons, the available archive images included TM images from Landsat 5 (July 1994 and June 2010). The obtained Landsat data were processed using the Level 1T Product Generation System (LPGS), which represents: GeoTIFF output format; cubic Convolution (CC) resampling method; 30-meter (TM) pixel size; Universal Transverse Mercator (UTM) map projection; World Geodetic System (WGS) 84 datum; MAP (North-up) image orientation. Level 1T included systematic radiometric and geometric accuracy by incorporating ground control points while employing a Digital Elevation Model (DEM) for topographic accuracy (DEM sources include SRTM, NED, CDED, DTED, and GTOPO 30) (http://landsat.usgs. gov/Landsat_Processing_Details.php, January, 2014). Other available Landsat images from USGS archives can be used to identify land cover changes in other vegetation seasons in an extended period: Landsat 8 OLI (September 2013); Landsat 7 ETM+ (March 2001 and April 2000); Landsat 5 TM (cloud cover 10 %, August 2007 and July 2007); Landsat 4 (July 1992) [31, 32].

3. The Area of Interest

The area of interest is located in the Tatra National Park, a part of the Carpathian Mountains, Figure 1. This is a unique example of alpine ecosystems in a comparatively small territory lacking any direct links to other alpine mountain ecosystems. Mountain forest plants are linked to a few vegetation zones. Spruce trees (*Picea abies*) are the dominant species in the montane zone (800-1500 meters above sea level). The exploration of land cover changes is based on the available archive images from Landsat scenes (from July 1994 to June 2010 for Landsat 5 TM, extended to September 2013 by Landsat 7 ETM+ and Landsat 8 OLI). The most important was the winter season in November 2004 when the territory was swept by gale-force winds with gusts of almost 200 km/hour (19 November 2004 between 3 p.m. and 8 p.m. local time). This storm caused the damage in the territory of the Tatra National Park (more than 12,000 hectares of forest growth was destroyed). Repeated forest fires followed several years after the calamity wind storm [33], Figure 2 and Figure 3. The damaged spruce trees are shown in Figure 4.

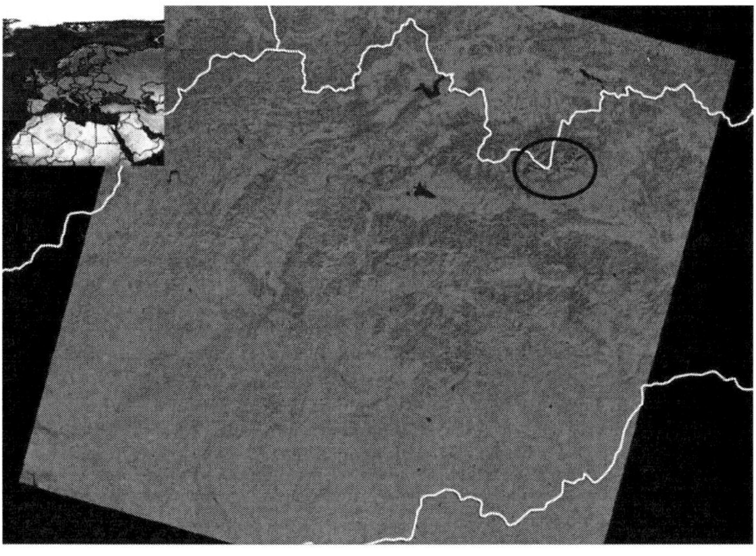

Figure 1. US Geological Survey (Global Visualization Viewer, Earth Resources Observation and Science Center: an overview of the whole area is shown in the upper left-hand part, the area of interest is marked by an ellipse on the Landsat 8 scene (ID: LC81880262013251LGN00, CC: 0%, Date: 2013/9/8, Qlty: 9, Product: OLI_TIRS_ L1T).

Figure 2. The area of interest in the Tatra National Park, the damage in a territory of the montane forest is represented by the sites situated to the south of the mountains (Landsat 8 OLI composite image, captured in September 2013, pixel size 30 meters, bands: 2, 3, 4).

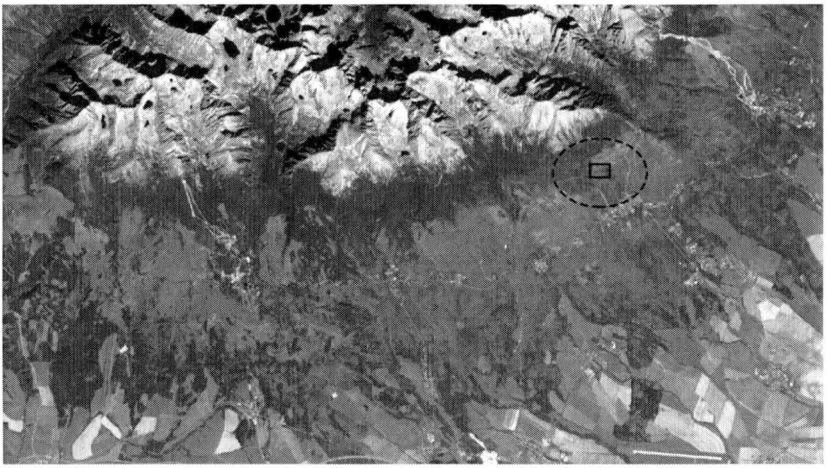

Figure 3. Close-up in the area of interest in the Tatra National Park, the damage in a territory of the montane forest is represented by the sites situated to the south of the mountains, a rectangle in the ellipse shows details of the sites disturbed by the wind storm (Landsat 8 OLI panchromatic image, captured in September 2013, pixel size 15 meters, band: 8).

In order to explore land cover changes for modelling vegetation dynamics and spatial patterns, unsupervised classification and supervised classification are used to detect land cover classes from image compositions based on Landsat 5 TM bands: band 1 blue-green (0.45-0.52); band 2 green (0.52-0.61); band 3 red (0.63-0.69); band 4 reflected IR (0.76-0.90); band 5 reflected IR (1.55-1.75); band 6 thermal (10.40-12.50); band 7 reflected IR (2.08-2.35). The selected area of interest is marked in Figure 5 and Figure 6 for the images captured in July 1994 and in June 2010, respectively.

The results of unsupervised classification for 4 classes are shown in Figure 7 (July 1994) and Figure 8 (June 2010). Based on the original images in Figure 5 and Figure 6, land cover classes are identified as: bare rock, spruce forest, disturbed forest site and agriculture areas. Land cover changes in a class spruce forest during an explored period are evident following the gale-force winds that swept the territory in November 2004. Distinguishing classes, such as areas disturbed by gale-force winds with subsequent fires and agricultural areas, is difficult due to the repeated human disturbances during the seasons of the year.

Taken from GoogleEarth, January 2014.

Figure 4. Detail of a site disturbed by the wind storm; thousands of hectares of the spruce forest were damaged in a similar way in the Tatra National Park.

The results of supervised classification for 3 training samples are shown in Figure 9 (July 1994) and Figure 10 (June 2010).

Figure 5. A composite image based on a Landsat 5 TM scene (July 1994, bands: 1, 2, 3); the selected area of interest for classification is marked on the right-hand side of the image by a rectangle.

Figure 6. A composite image based on a Landsat 5 TM scene (June 2010, bands: 1, 2, 3); the selected area of interest for classification is marked on the right-hand side of the image by a rectangle.

Figure 7. An unsupervised classification based on the image from a Landsat 5 TM scene (July 1994); 4 land cover classes in comparison with the original composite images are identified as: bare rock, spruce forest, disturbed forest sites and agriculture areas.

Figure 8. An unsupervised classification based on the image from a Landsat 5 TM scene (June 2010); 4 land cover classes in comparison with the original composite image are identified as: bare rock, spruce forest, disturbed forest sites and agriculture areas.

Figure 9. A supervised classification for 3 training samples based on the image from a Landsat 5 TM scene (July 1994); 3 land cover classes in comparison with the original composite image are identified by training samples as: bare rock, spruce forest, disturbed forest sites and agriculture areas.

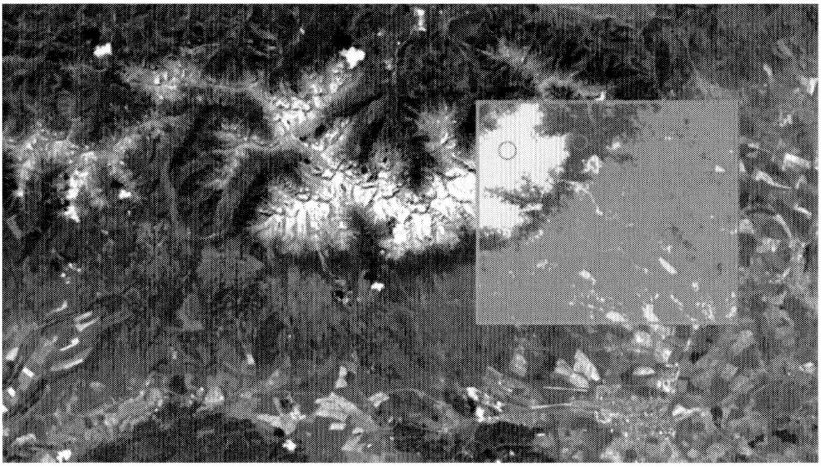

Figure 10. A supervised classification for 3 training samples based on the image from a Landsat 5 TM scene (June 2010); 3 land cover classes in comparison with the original composite image are identified by training samples as: bare rock, spruce forest, disturbed forest sites and agriculture areas.

In comparison with the previous unsupervised classification, the number of classes is decreased to clearly determine the remainder of the spruce forest and the sites disturbed by strong winds, together with agricultural areas.

Training samples for supervised classification are marked by circles.

Each sample includes representative pixels that are attached in advance to a determined land cover class.

Other pixels in the area of interest are identified by multivariate statistics on the basis of similar optical properties. Due to the different approach, the results of unsupervised and supervised classification can be different in dependence on setting of the input parameters.

4. Modeling Approches for Vegetation Dynamics and Spatial Patterns

Changes in mountain forest conditions are not related to a single factor but rather to a mixture of stress factors that intensified one another [34, 35]. The gale-force winds with gusts of almost 200 km/hour in November 2004 were the most important factor in the explored period from 1994 to 2010. Together with repeated forest fires that followed several years after the wind storm calamity, they caused the damage in the territory.

In order to describe subsequent forest regeneration and growth, the relevant processes need to be quantified by modelling on detailed spatial and temporal scales. A number of studies have been carried out on tree growth and regeneration in closed forests and in windthrow areas [36-38].

A widely used group of models represents gap models that can also examine the effects of climate change on forest development. The lack of appropriate empirical data focused on the spatial and temporal extent of forest succession renders model evaluation a non-trivial task [39-42].

A new approach is represented by spatio-temporal models that extend geographic information systems (GISs) to provide spatial analysis together with the development of dynamic modelling tools [43-46].

Natural regeneration of *Picea abies* in the disturbed sites can be predicted by two basic phases: competition among seedlings (*Picea abies*) and ground vegetation [47]; competition between small trees (*Picea abies*) and other trees.

The survival of seedlings depends on the local conditions of individual microsites. Ingrowth into the seedlings is assumed to be a function of viable seeds, microsite-specific germination rate and the mortality rate until the first summer.

The number of seeds for model predictions can be generated randomly in dependence on the density of local trees, or equated to the measured value for a chosen seed year. Spruce seedling survival can be described as:

$$N_{(x,y,k)} = N_{0(x,y,k)} \cdot f_{random(x,y,k)} \tag{1}$$

$$\frac{dN_{(x,y,k)}}{dt} = r_{(x,y,k)} \cdot N_{(x,y,k)}. \tag{2}$$

The ordinary differential equations are processed for each age group *k* of a number of seedlings $N_{x,y,k}$ in each microsite. The initial number of seedlings $N_{0x,y,k}$ and the growth rate $r_{x,y,k}$ are dependent on a microsite location [x, y], seasonal conditions and the specific age group *k*. The random function $f_{random\_x,y,k}$ is used to generate seedlings in dependence on seed years (1 for a seed year, 0 for no seed year).

Due to the high mortality of seedlings, the occurrence of individual small trees for the next phase is rather rare. As some trees grow faster in dependence on local conditions, the conversion should prefer size-based values to age-based values. The growth predictions for individual trees can be based on estimates originating from previous observation of individual sites in the area of interest. The growth of individual trees is assumed to be independent of the microsite ground vegetation they inhabit. The system of tree survival and diameter models can be generally described as:

$$p_{i,k} = f_{probability(x,y,k)} \tag{3}$$

$$\Delta d_{i,k} = d_i f_{d(x,y,k)}, \tag{4}$$

where the probability of survival $p_{i,k}$ and the increment of the tree diameter $\Delta d_{i,k}$ are dependent on the local environment of the site at a location [x, y]. Parameter *i* is the tree identifier. The parameter *k* marks a tree age group.

GIS can be used to manage spatial data from image classification, estimate model parameters from a series of classified images, provide pre-processing and post-processing of data for numerical modelling, and display the results. For the described area of interest in the Tatra National Park and examined period from July 1994 to June 2010, there were two available Landsat 5 TM images in the USGS archive. Thus, it is possible to assess the forest land cover only 10 years prior to the wind storm and 6 years after the storm in the explored area.

The target data structures for numerical solution of the model are formed by grids of cells. The size of a cell in the grid is dependent on the size of a pixel in the classified images. For the Landsat 5 TM images, the size of a pixel is 30 meters.

In dependence on the size of the whole area of interest, the minimum cell size can be 30 meters. The numerical solution of a spatio-temporal model for seedlings of *Picea abies* is based on a set of differential equations for the grid of cells. The growth predictions for the trees are based on a set of algebraic equations for each tree.

All the equations can be solved by GIS or by a standalone computer program. The numerical solution can be included in the GIS as a series of thematic map layers that include the actual state of the spruce forest in selected time points. At the present time, GIS can be used for nearly all tasks focused on spatio-temporal modelling. For time-consuming numerical simulations, it is necessary to couple GIS with advanced software tools designed for high performance in dynamic modelling. However, the increasing power of GIS tools enables reconsideration of this relationship.

CONCLUSION

The area of interest in the Tatra National Park, part of the Carpathian Mountains, represents an example of an alpine ecosystem in a comparatively small territory, but the model can be applied to other alpine mountain ecosystems or to other similar sites in the world.

Despite a lack of image data for classification in specific seasons of the year in the explored period (1994 2010), the archived Landsat data provided a highly valuable source of information. Programs focused on remote sensing, such as the Landsat Program, can provide information that is very useful for analysing land cover changes in the mountain forest environment.

The potentials and limits require far more exploration and refinement, however, and this process is currently under way. The attached unsupervised and supervised classifications demonstrate processing of multispectral images from Landsat satellites. Proposal of a spatio-temporal model enables us to obtain predictions of regeneration and growth of a spruce forest after a storm. GIS can integrate all the spatial data and predictions. It can be a valuable foundation for decision-making processes in natural conservation and risk assessment analysis.

Acknowledgments

The described research was carried out in the framework of the GIS Laboratory research project supported by the Ministry of Education, Youth and Sports, the PRVOUK-environmental research project and partially by the GACR 526-09-0567 project. The ArcGIS and MATLAB software tools were used for processing of the presented images and development of the model.

References

[1] Botkin, D. B. (1993). *Forest dynamics: an ecological model*. New York: Oxford University Press.

[2] Alaric, S. V. ed. (1994). *Remote Sensing and GIS in ecosystem management*. Washington: Island Press.

[3] Hunziker, U., Brang, P. (2005). Microsite patterns of conifer seedling establishment and growth in a mixed stand in the southern Alps. *Forest Ecology and Management* 162, 67–79.

[4] Kolström, M. (1998). Ecological simulation model for studying diversity of stand structure in boreal forests. *Ecological Modelling* 111, 17–36.

[5] Sterba, H., Monserud, R. A., 1997. Applicability of the forest stand growth simulator PROGNAUS for the Austrian part of the Bohemian Massif. *Ecological Modelling* 98, 23–34.

[6] Campbell, J. B. (2002). *Introduction to Remote Sensing*. 3rd edition. New York: Taylor and Francis.

[7] Franklin, S. E. (2001). *Remote Sensing for Sustainable Forest Management*. London: CRC Press.

[8] Boyd, D. S. and Foody, G. M. (2011). An overview of recent remote sensing and GIS based research in ecological informatics. *Ecological Informatics 6*, 25-36.

[9] Short, N. M. (1986). *Geomorphology from Space: A Global Overview of Regional Landforms*. United States: Government Printing.

[10] Canadian Council of Forest Ministers (1995). *Defining Sustainable Forest Management. A Canadian Approach to Criteria and Indicators.* Ottawa: Canadian Forest Service, Natural Resources.

[11] Wolter, P. T., Townsend, P. A., Sturtevant, B. R., and Kingdon (2008). Remote sensing of the distribution and abundance of host species for

spruce budworm in Northern Minnesota and Ontario. *Remote Sensing of Environment* 112, 3971-3982.

[12] Meroni, M., Panigada, C., Rossini, M., Picchi, V., Cogliati, S., and Colombo, R. (2011). Using optical remote sensing techniques to track the development of ozone-induced stress. *Environmental Pollution* 157, 1413-1420.

[13] Wimberly, M. C. and Reilly, M. J. (2007). Assessment of fire severity and species diversity in the southern Appalachians using Landsat TM and ETM+ imagery. *Remote Sensing of Environment* 108, 189-197.

[14] Meng, Q., Cieszewski, C. and Madden, M. (2009). Large area forest inventory using Landsat ETM+: A geostatistical approach. *ISPRS Journal of Photogrammetry and Remote Sensing* 64, 27-36.

[15] Van Wagtendonk, J. W., Root, R. R. and Key, C. H. (2004). Comparison of AVIRIS and Landsat ETM+ detection capabilities for burn severity. *Remote Sensing of Environment* 92, 397-408.

[16] Chen, J., Zhu, X., Vogelmann, J. E., Gao, F., and Jin, S. (2011). A simple and effective method for filling gaps in Landsat ETM+ SLC-off images. *Remote Sensing of Environment* 115, 1053-1064.

[18] Pôças, I., Cunha, M. and Pereira, L. S. (2011). Remote sensing based indicators of changes in a mountain rural landscape of Northeast Portugal. *Applied Geography* 31, 871-880.

[19] Mulders, M. A. (2001). Advances in the application of remote sensing and for surveying mountainous land. *Remote Sensing and GIS for Surveying Mountainous land* 3(1), 1-8.

[20] Metternicht, G., Hurni, L. and Gogu, R. (2005). Remote sensing of landslides: An analysis of the potential contribution to geo-spatial systems for hazard assessment in mountainous environments. *Remote Sensing of Environment* 98, 284-303.

[21] Mantovani, F., Soeters, R. and Van Westen, C. J. (1996). Remote sensing techniques for landslides studies and hazard zonation in Europe. *Geomorphology* 15, 213-225.

[22] Mulder, V. L., de Bruin, S., Schaepman, M. E., and Mayr, T. R. (2011). The use of remote sensing in soil and terrain mapping-A review. *Geoderma* 162, 1-19.

[23] Kovalskyy, V. and Roy, D. P. (2013). The global availability of Landsat 5 TM and Landsat 7 ETM+ land surface observations and implications for global 30 m Landsat data product generation. *Remote Sensing of Environment* 130, 280-293.

[24] Markham, B. L. and Helder, D. L. (2012). Forty-year calibrated record of earth-reflected radiance from Landsat: A review. *Remote Sensing of Environment* 122, 30-40.

[25] Loveland, T. R. and Dwyer, J. L. (2012). Landsat: Building a strong future. *Remote Sensing of Environment* 122, 22-29.

[26] Wulder, M. A., White, J. C., Goward, S. N., Masek, J. G. Irons, J. R., Herold, M., Cohen, W. B., Loveland, T. R., and Woodcock, C. E. (2008). Landsat continuity: Issues and opportunities for land cover monitoring. *Remote Sensing of Environment* 112, 955-969.

[27] Jensen, J. R. (2004). *Introductory Digital Image Processing: A Remote Sensing Perspective*. 3rd edition. New York: Prentice Hall.

[28] Johnson, R. A. and Dean, W. W. (2007). *Applied Multivariate Statistical Analysis*. 6th edition. New York: Prentice Hall.

[29] Richards, J. A. (2013). *Remote Sensing Digital Image Analysis: An Introduction*. 5th edition. New York: Prentice Hall.

[30] Tso, B. and Mather, P. M. (2001). *Classification Methods for Remotely Sensed Data*. London: Taylor and Francis.

[31] Kovalskyy, V. and Roy, D. P. (2013). The global availability of Landsat 5 TM and Landsat 7 ETM+ land surface observations and implications for global 30 m Landsat data product generation. *Remote Sensing of Environment* 130, 280-293.

[32] Gutman, G., Huang, C., Chander, G., Noojipady, P., and Masek, J. G. (2013). Assessment of the NASA–USGS Global Land Survey (GLS) datasets. *Remote Sensing of Environment* 134, 249-265.

[33] Crofts, R., Zupancic-Vicar, M., Marghescu, T., and Tederko, Z. 2005. IUCN Mission to Tatra Natural Park, Slovakia, April 2005. IUCN-the World Conservation Union (http://www.wolf.sk/files/dokumenty/IUCN_EN_zaverecna_sprava_2005.pdf January 2014).

[34] Spiecker, H. ed. (2004). *Norway Spruce conversion-option and consequences*. Boston: Brill.

[35] Forman, R. T. T. (1995). *Land Mosaics: The Ecology of Landscape and Regions*. Cambridge: Cambridge University Press.

[36] Liu, J., Ashton, P. S. (1995). Individual-based simulation models for forest succession and management. *Forest Ecology and Management* 73, 157–175.

[37] Didion, M., Kupferschmid, A. D., Lexer, M. J., Rammer, W., Seidl, R., Bugmann, H., (2009). Potentials and limitations of using large-scale forest inventory data for evaluating forest succession models. *Ecological Modelling* 220, 133–147.

[38] Gudrun, W., Tappeiner, U., Strobl, J., and Tasser, E. (2008). Understanding alpine tree line dynamics: An individual-based model. *Ecological Modelling* 218, 235-246.
[39] Meer, P. J., Jorritsma, I. T. M., Kramer, K. (2002). Assessing climate change effects on long-term forest development: adjusting growth, phenology, and seed production in a gap model. *Forest Ecology and Management* 162, 39–52.
[40] Hanssen, K. H. (2003). Natural regeneration of Picea abies on small clear-cuts in SE Norway. *Forest Ecology and Management* 180, 199–213.
[41] Risch, A. C., Heiri, C., Bugmann, H. (2005). Simulating structural forest patterns with a forest gap model: a model evaluation. *Ecological Modelling* 181, 161–172.
[42] Pennanen, J., Greene, D. F., Fortin, M. J., and Messier C. (2004). Spatially explicit simulation of long-term boreal forest landscape dynamics: incorporation quantitative stand attributes. *Ecological Modelling* 180, 195-209.
[43] Goodchild, M. F., Steyaert, L. T. and Parks, B. O. (1996). *GIS and Environmental Modeling: Progress and Research Issues*. Fort Collins: GIS World, Inc.
[44] Matejicek, L. (2012). *Environmental Modeling with GIS*. New York: Nova Science Publishers.
[45] Qiu, F., Li, B., Chastain, B., and Alfarhan, M. (2008). A GIS based spatially explicit model of dispersal agent behavior. *Ecological Modelling* 254, 524-537.
[46] Vogelmann, J. E., Xian, G., Homer, C., and Tolk, B. (2012). Monitoring gradual ecosystem change using Landsat time series analyses: Case studies in selected forest and rangeland ecosystems. *Remote Sensing of Environment* 122, 92-105.
[47] Matejicek, L., Vavrova, E. and Cudlin P. (2011). Spatio-temporal modelling of ground vegetation development in mountain spruce forests. *Ecological Modelling* 222, 2584-2592.

In: Forest Ecosystems
Editor: Noel C. Roberts
ISBN: 978-1-63117-815-3

Chapter 3

THE ROLE OF SPRUCE IN RED SQUIRREL (*SCIURUS VULGARIS*) CONSERVATION IN THE UK

***P. W. W. Lurz*[1,*] *and C. M. Shuttleworth*[2]**
[1]Royal (Dick) School of Veterinary Studies,
Edinburgh, United Kingdom
[2]School of Environment, Natural Resources and Geography,
Bangor University, Bangor, Gwynedd,
United Kingdom

ABSTRACT

Spruce dominated forests undoubtedly represent one of the conservation ironies of the 21st century in the UK. These introduced tree species are planted extensively in commercial forestry and now offer a refuge to the endangered native Eurasian red squirrel (Sciurus vulgaris) in the face of competition from the introduced North American grey squirrel (S. carolinensis). The present chapter provides a concise historical overview of red squirrel population decline within the UK, describes the respective ecology and competitive interactions of both red and grey squirrels and the subsequent evolution of current conservation strategies. The importance of spruce dominated forests in the conservation of red squirrels is critically reviewed; population dynamics and habitat use are

* Email: lurzpww@gmail.com

explored with particular focus upon patterns of tree seed availability, and the integration of squirrel conservation with parallel commercial forest management targets are described. Knowledge gaps in the proactive management of remnant red squirrel populations are presented, and we explore the potential impact of emerging disease threats to both squirrels and UK forest ecosystems.

INTRODUCTION

The plight, and near complete disappearance, of the native Eurasian red squirrel (*Sciurus vulgaris*) (Figure 1) over the last 80 years from most of the UK is almost singularly the result of the introduction of the North American grey squirrel (*S. carolinensis*). This ecological replacement has been reported in great detail in the scientific literature and the media [e.g., 1-4]. What has perhaps received less attention is that spruce dominated, upland conifer plantations in the North of England, Wales and Scotland (Figure 2) are seen as crucial stronghold areas for red squirrel conservation and the survival of the species in the UK. It is undoubtedly a conservation irony, that forest habitats dominated by introduced tree species such as Sitka (*Picea sitchensis*) or Norway spruce (*P. abies*) are seen as a last refuge for an endangered native mammal threatened by a highly successful non-native competitor.

Figure 1. Eurasian red squirrel.

Figure 2. Sitka spruce dominated forest in northern England (Photo: P. Lurz).

The grey squirrel is included as an example in the world's worst 100 alien invasive species compiled by the IUCN [5], and in Europe it has unfortunately been introduced to Britain, Ireland and Italy. Grey squirrels not only compete for resources with the red squirrel, but carry squirrelpox virus as an asymptomatic infection; a virus that produces pathogenic disease in their congener [6-8]. Additionally, the grey squirrel is considered a significant forest pest due to its bark-stripping habits [9, 10]. Whilst interactions between the two squirrel species, on the face of it, may appear simple with the native being replaced by the introduced species, the mechanisms of this process are often complex, and are influenced by food availability, forest composition, how fragmented (or connected) forest landscapes are, squirrel density and the presence of disease [11-14].

Grey squirrels are broadleaf specialists, and in their native range inhabit deciduous forests in the eastern parts of the Unites States. They are most common in woodlands with a diverse woody understory containing large seeded tree species such as oak (*Quercus*), hickory (*Carya*) or walnut (*Juglans*) species [15]. They have a significant competitive edge over red squirrels in deciduous woodlands in the UK, particularly where oak (Quercus spp.) is present [9]. In contrast, grey squirrels have high (65%) population turnover in conifer habitat [16] with low juvenile survival rates [17]. Kenward et al. [18] postulated that the higher mortality rates observed in conifer relative to broadleaved habitats would mean that population persistence would be reliant upon continual immigration from adjacent broadleaves. Sightings data recorded for red squirrels in the UK also show that remnant populations have survived for longest in conifer forests in areas where grey squirrels were present [19]. Spruce-dominated commercial forest plantations therefore represent a key red squirrel conservation focus.

The concept of conifer dominated conservation areas for red squirrels was pioneered by Harry Pepper and John Gurnell [20, 21], and the first refuge was officially designated in 1995 at Spadeadam Forest a large, spruce-dominated conifer forest in Northern England. This coincided with the publication of the Biodiversity Action Plan and UK strategy for red squirrel conservation [22]. In Northern England, the Forestry Commission and People's Trust for Endangered Species (PTES) supported scientific research to develop and implement management guidelines for red squirrels in Sitka spruce dominated plantations. This led to the designation of stands within Kielder, Kidland and Usway forests as red squirrel reserves in 2003 [22, 23]. This conservation network was expanded three years later to encompass 16 squirrel reserves in the North of England [22]. In total, 17 conifer forest areas have now been designated in England, with an additional 18 proposed in Scotland as part of national landscape scale strategies to conserve the remaining red squirrel populations [22, 24, 25].

Unique Character of Spruce Plantations

Commercial spruce dominated forests are often planted on exposed and wind-swept upland sites with associated nutrient poor, acid and wet soils, that limit both the selection of tree species for planting and the subsequent management of forest stand and age structures. Whilst the public commonly perceive forests as a static and permanent landscape feature, upland commercial spruce forests and their associated mammal populations are in fact in constant change. High wind-throw risks on exposed hill sides typically lead to crop rotation times of less than 50 years; and at the forest scale there is an on going process of harvesting and replanting with resulting shifts in forest composition and age classes of stands. This scenario poses unique challenges for red squirrel population management as harvesting and restocking processes alter local habitat suitability for squirrels and other animals and consequently how many individuals the forest resource can support.

An additional factor is that spruce crops, such as Sitka, Norway or Serbian spruce (*P. omorika*) tend to be planted in single species blocks known as coupes, rather than in the intimate species mixtures found in natural woodlands. This typically creates a large-scale patchwork mosaic across the wider landscape, with alternating blocks of different crop species interspersed with open areas where timber harvesting has taken place (Figure 2). Tree species differ in their suitability for squirrels due to relative differences in seed size, nutritional quality, episodic patterns of coning and seasonal differences in seed shed. Annual autumn fallen cone counts on transect lines within Sitka and Norway spruce plantations in Spadeadam and Kidland Forests (Northern England) offer a good illustration of the potential annual fluctuations in seed crops; the main diet of red squirrels [26-29]. The estimated cone crop based on transect lines for Norway spruce at Spadeadam Forest showed a 34-fold difference between 1991 and 1992 (Figure 3); and data for Sitka spruce indicated a 17.8 and 9.3-fold difference between the lowest and highest cone crops measured at Spadeadam (1991-1999) and Kidland Forest (2001-2013) respectively.

Intervals between poor and good (mast) seed crops vary with tree species and are influenced by local conditions such as soils, climate and pollution [19, 30]. Reported cycles between poor and good crops in the literature suggest 3-5 years for Sitka and 6-7 or up to 10-11 years for Norway spruce [31-34], but see also shorter 4-year cycles reported by [35]. As seed quality in terms of protein and crude fat contents can also vary significantly between years [36], red squirrels face considerable and sometime dramatic variation in food availability in spruce plantations.

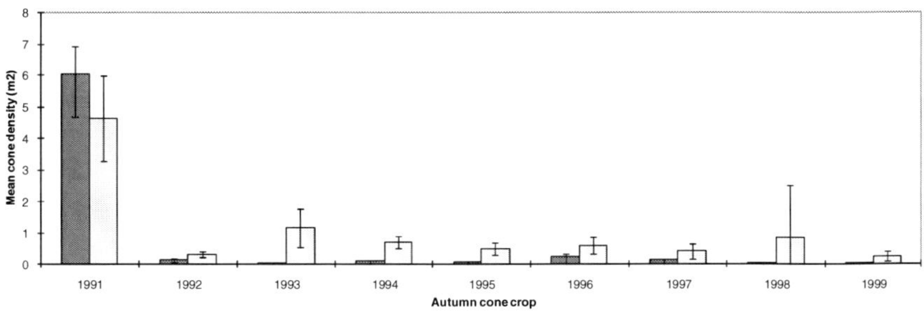

Figure 3. Annual autumn cone crop measurements (cones $m^{-2} \pm$ 1STD) for Norway (dark bars) and Sitka spruce (light bars) at Spadeadam Forest, Northern England between 1991-1999. The relatively large standard deviations indicate the patchiness of cones in trees across the forest.

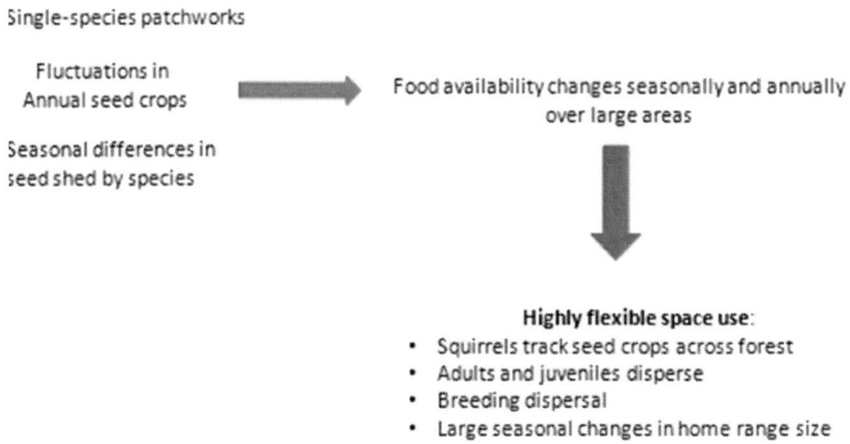

Figure 4. Diagram illustrating factors influencing red squirrel space use behavior in spruce dominated plantations.

These changes in food availability (and quality) have direct impacts on red squirrel numbers, productivity, adult survival as well as the survival of any offspring produced [29, 37]. Health surveillance studies have confirmed starvation as a cause of mortality in sub-adult animals [38]. If resident in an extensive monoculture spruce stand, a red squirrel with an adult life-expectancy of 3-5 years would expect to encounter one large cone crop in Sitka spruce, or have a 50% chance of encountering one large crop in Norway spruce during its life. As a result, large-scale single species stands combined with drastic annual changes in food supply (seed crops) have marked consequences on squirrel residency, space use, social behavior and settlement patterns. In monoculture conifer stands residency rates in the face of seed crop failures can be low, with 31% recorded in Corsican pine (*Pinus nigra*) [39] and as low as 0% in monoculture Sitka spruce [26].

In contrast to the relatively stable red squirrel social organization often found within mixed deciduous habitats with a more dependable seed source [40], male red squirrels in large spruce plantations appear to track females and females track spatially changing seed crops across the forest landscape mosaic [29]. Home range sizes are very large for an animal that weighs less than 400g (Average of 6.29 ha for females and 17.8 ha for males in spruce dominated forests) and reflect the patchiness of food availability. Seasonal adjustments to home range size, or the complete abandonment of previous home ranges and breeding dispersal, whereby adult females bequeath part or all of their home range to their offspring and disperse instead of the juvenile, were typical features of this dynamic landscape [26, 29, 41]. Figure 4 illustrates the factors that shape the highly adaptive and flexible squirrel space use in these spruce dominated forests.

Nest Use

Data for Sitka, Norway and Serbian spruce indicate that all three species are used for the location of squirrel nests (termed dreys). In fact, squirrels showed a preference for spruce canopies and the increased shelter their denser crown structures offer over neighboring pine species with their more open canopies [26]. Dreys in spruce are usually located several meters up within the lower part of the crown next to the main stem. Most squirrels have access to more than one drey at any given time and radio-tracking data at Spadeadam Forest suggested an average of 3.44 (range of 1-8), some of which were also used by other squirrels at other times. Drey construction is an intense but relatively low-cost effort. A project using a hidden camera inside a nestbox filming a free-ranging red squirrel building a complete drey in February 2012 in Germany logged 3.6 hours of construction inside the nestbox (excluding collection of materials outside) over a period of 4 days [42].

Grey Squirrels in Spruce Plantations

Adult red squirrel survival rates, densities and female breeding rates in coniferous woodlands remain relatively unaffected in the presence of grey squirrels [11]. However, juvenile recruitment was reduced from 50% to 13%, and red squirrel ranges tended to be focused upon spruce rather than pine stands in Northern England and Italy [11]. In a subsequent Scottish conifer forest study, grey squirrels selected strongly for mixed conifers and broadleaves in all seasons, while Sitka spruce, Norway spruce and Douglas fir (*Pseudotsuga menziesii)* were rarely used [43]. Sitka spruce appears to be a suboptimal habitat type for grey squirrels [44, 45], and it has been suggested that as grey squirrel densities increase, sympatric red squirrel populations, would progressively be forced into these habitats [44]. Preliminary computer models on limited data [46] have suggested that in such a scenario, ultimately extinction could occur in all but the most extensive spruce dominated habitats.

One of those habitats modelled was the 4700ha spruce dominated Clocaenog forest in North Wales. In the late 1990s, red squirrel densities (0.28-0.30 animals per hectare) were similar to sympatric grey squirrel densities (0.24-0.30) [45]. Subsequent grey squirrel control in the forest was localized with no trapping taking place in some years, and with productive adjacent broadleaved valley populations also left largely uncontrolled. Winter trapping in 2011/12 within five Sitka spruce dominated forest areas removed 144 grey squirrels with densities of 0.08 to 0.37 animals per hectare, with only one parallel red squirrel sighting recorded [47]. The study also measured grey squirrel body fat content and found some animals in very good condition (Figure 5); these individuals potentially benefited from the presence of oak (*Q. petraea*), beech (*Fagus sylvatica*), and pine species within the forest.

There still, however, remains a paucity of information on the temporal and spatial patterns of habitat use in spruce plantations by grey squirrels, and whether, in the absence of culling, such habitats represent a long-term red squirrel refuge. The presence of animals with high body reserves observed in Clocaenog also suggests that populations may have a permanent presence within spruce forests containing oak, beech and pine, and this scenario would have implications for disease transmission.

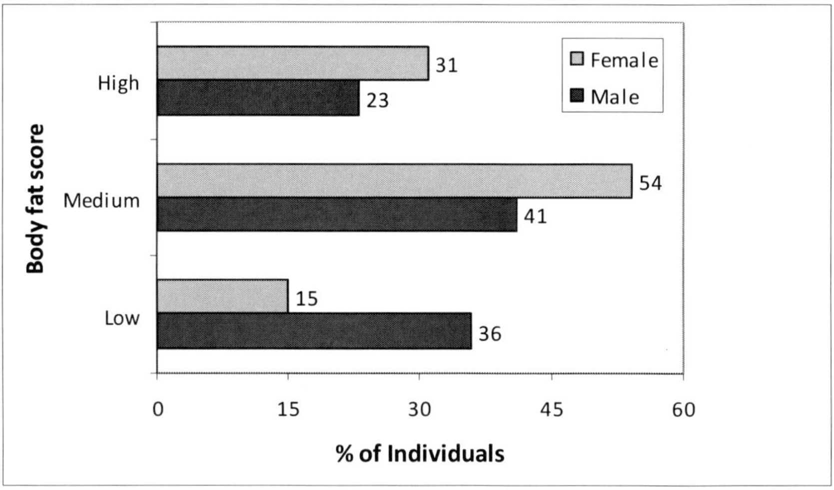

Figure 5. Body fat scores of grey squirrels killed in spruce dominated stands during the winter 2011/12 [47].

DISEASE

Determining causes of mortality in upland spruce is difficult due to the low population densities. Nest box studies have yielded carcasses, but these are found infrequently and are generally unsuitable for post mortem due to advanced decomposition [48]. Despite these challenges, using infection data derived from culled grey squirrels and projections generated in computer modelling, it is possible to quantify viral infection rates and implications for pathogenic disease. Recent research has focused upon two viral types: adenovirus and squirrelpox virus.

Adenovirus can produce pathogenic disease and mortality clusters in wild red squirrels [49, 50]. The virus is also carried asymptomatically as a latent or subclinical infection in red squirrels, and is found in grey squirrels [50, 51] and woodmice (*Apodemus sylvaticus*) [52], but the importance of inter-specific infection pathways remain opaque. It is likely that low densities of greys in spruce would minimize the associated risk of cross infection relative to pine and broadleaved habitats where densities are higher [18, 19]. Given the fact that squirrelpox can produce epidemic disease in red squirrels, of perhaps greater significance is the impact that carrying capacity may have upon infection rates in resident grey squirrels.

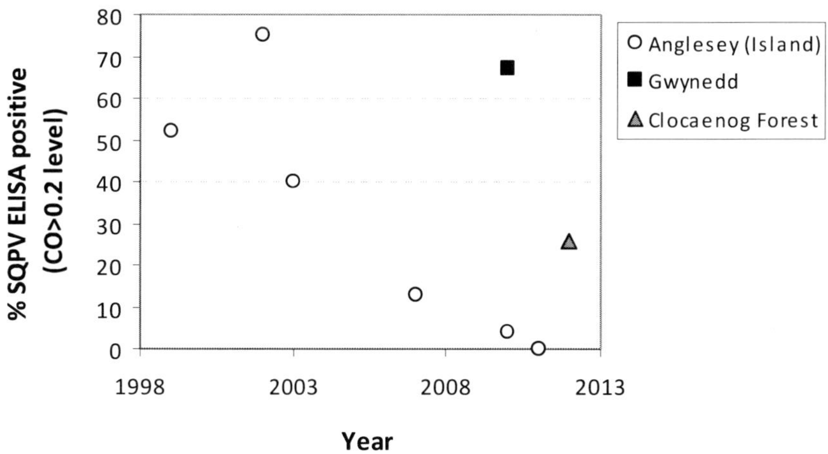

Figure 6. The percentage of adult grey squirrel blood tests positive for squirrel pox virus antibodies. Data are from mixed deciduous and pine forest habitats on Anglesey [47, 53], Gwynedd broadleaved and mixed coniferous habitats and Clocaenog Sitka spruce stands. Gwynedd sample n=70, Clocaenog n=129, Anglesey n=23-99 (1999-2010) and n=2 (2011).

It is now known that squirrelpox infection rates can decrease in broadleaved grey squirrel populations exposed to sustained and intensive culling. Observed rates of antibody response determined by enzyme-linked immunosorbent assay (ELISA) declined within Anglesey populations during their eradication. Initially in 1999-2003 >50% of adults tested positive to antibodies, but this fell to 4% by 2010; a pattern which continued to 0% reported in 2011 when greys were close to extinction on the island [47]. In the adjacent mainland (Gwynedd county) mixed deciduous and pine woodland, antibodies were found in almost 70% of animals tested (Figure 6). Interestingly, in Clocaenog forest during 2012, overall rates were less than 30%, which may reflect low inter-animal encounter frequencies within an established but low density population.

A recent computer modelling approach [53] based on extensive fieldwork, culling and antibody blood sampling data from Scotland, highlighted the importance of managing disease threat not just in terms of removal of animals spreading the virus, but also in terms of population density and thus habitat quality or carrying capacity. Predictions showed that low densities of squirrels, such as those found in spruce dominated plantations, restricted both disease epidemics and disease spread. Regional disease outbreaks in red squirrel populations have caused populations crashes making the species more vulnerable to replacement and extinction by sympatric grey squirrels. This raises important questions for the management of the established UK network of spruce dominated strongholds and suggests that maximizing red squirrel population density through habitat measures (e.g., planting only tree species that support high densities or by supplementary feeding) can increase disease risk and should be avoided. The simulations also illustrated the difficulty of successfully preventing disease spread in landscapes with high levels of woodland cover and connectivity; and indicated the importance of dispersal 'pinch points' for limiting disease spread.

Future Conservation of Red Squirrels in Spruce-Dominated Forests

Current management strategies seek to reduce the food energy resource available for grey squirrels by selecting stronghold forest areas dominated by conifer species such as Sitka spruce that produce a small sized seed [28]. This makes it challenging for grey squirrels to obtain their daily requirements, and minimizes habitat attractiveness; thereby limiting the competitive and disease risk they pose to native red squirrels.

In turn, forest management plans aim to provide a diverse, dependable food supply for red squirrels through careful proactive planning of forest composition and age structure. Recommendations favor the planting of tree and shrub species that help red squirrels, or are neutral and so avoid giving grey squirrels a competitive advantage. Planning also includes measures to reduce any disturbance caused by forest operations, such as through the provision of arboreal corridors for dispersal during timber harvesting and clear cutting [54]. Potential conflicts with other strategic objectives can be minimized by managing the forest stronghold as an interlinked woodland system which flexibly maintains mature seed-bearing forest stands to support a minimum viable population [54]. However, uncertainties in the form of climatic change and the emergence of pathogenic tree diseases increasingly present challenges to established red squirrel conservation approaches [55].

For example, many of the commercial pine species planted as a component in upland spruce dominated plantations are now vulnerable to defoliation by Red Band Needle Blight (*Dothistroma septosporum*) [56]. Similarly, the pathogen *Phythopthora ramorum,* which infects a wide range of trees, has been found to infect and kill larch (*Larix* spp.) trees in the UK and planting restrictions for larch have been imposed in some forests [57]. Tree diseases therefore severely restrict the range of tree species available for re-stocking and hence the future provision of a varied food supply throughout the year in forests managed for red squirrels. The threat of grey squirrels further precludes the selection of alternative native crop species (e.g., oak). And whilst some alternative non-native conifer species have been advocated as drought-resistant alternatives for conifer forests in the UK at risk from climate change [58], there are few ecological studies that have assessed the impact of these species on native vertebrates in general; and none that focused on their effects on red-grey squirrel interactions specifically [55].

Spruce forests have offered a last refuge for red squirrels in the UK. However, there is currently a knowledge gap on how best to maintain suitable forest habitats for this mammal and other native species in the face of future climate change and emerging diseases. The relative safety and current reliance upon Sitka spruce in red squirrel conservation represents a vulnerability should a pathogen emerge that adversely affects this species. We believe that there is a need for research on the suitability and likely impacts of proposed future commercial tree species on ecosystem health, function and potential ecosystem services (conservation benefits).

ACKNOWLEDGMENTS

The authors would like to thank Forestry Commission, Kielder Forest District and in particular Neville Geddes and Bill Burlton. We extend our gratitude to the People's Trust for Endangered Species, Red Squirrels Trust Wales and Natural Resources Wales for their support of red squirrel research over the years.

REFERENCES

[1] A. D. Middleton, *Proc. Zool. Soc. Lond.* 1930, 809-843 (1930).
[2] H. G. Lloyd. *Mamm. Rev*. 13, 69-80 (1983).
[3] J. Bryce. *Mamm. Rev*. 27, 171-176 (1997).
[4] P. W. W. Lurz. Red Squirrel. Naturally Scottish Series, SNH, Edinburgh (2010).
[5] S. Lowe, M. Browne, Boudjelas, S. and M. De Poorter. *100 of the World's Worst Invasive Alien Species. A Selection from the Global Invasive Species Database.* ISSG, World Conservation Union (IUCN), Gland, (2000).
[6] S. P. Rushton, P. W. W. Lurz, J. Gurnell, P. Nettleton, C. Bruemmer, M. D. F. Shirley and A. W. Sainsbury. *Epid. & Inf.* 134, 521-533 (2006).
[7] A. W. Sainsbury, P. Nettleton, J. Gilray and J. Gurnell. *Anim. Cons*. 3, 229-233 (2000).
[8] A. W. Sainsbury, R. Deaville, B. Larsen, W. A. Cooley, S. S. Farrelly, M. J. Stack, P. Duff, C. J. McInnes, J. Gurnell, P. H. Russel, S. P. Rushton, D. U. Pfeiffer, P. Nettleton and P. W. W. Lurz. *Ecohealth* 5, 305-316.
[9] R. E. Kenward. *Mamm. Rev*. 13, 159-166 (1983).
[10] L. Huxley. *The Grey Squirrel Review*. European Squirrel Initiative, Suffolk, UK (2003).
[11] L. A. Wauters, P. W. W. Lurz and J. Gurnell. *Ecol. Res*. 15, 271-284 (2000).
[12] L. A. Wauters, G. Tosi and J. Gurnell. *Behav. Ecol. and Sociobiol*. 51, 360-367 (2002).
[13] J. Gurnell, L. A. Wauters, P. W. W. Lurz and G. Tosi. *J. Anim. Ecol*. 73, 26-35 (2004).

[14] S. Bosch and P. W. W. Lurz. *Eurasian Red Squirrel.* Westarp Wissenschaften Verlagsgesellschaft, Germany (2012).
[15] J. L. Koprowski. *Mamm. Species* 480, 1-9 (1994).
[16] P. W. W. Lurz and A. J. Lloyd. *J. Zool., Lond.* 252, 539-543 (2000).
[17] D. Smith and J. Gurnell. The ecology of the grey squirrel in conifer forest. In: Gurnell J and Lurz PWW, editors. *The conservation of red squirrels, Sciurus vulgaris* L.. PTES; London; 1997; pp 109-120.
[18] R. E. Kenward, K. H. Hodder, R. J. Rose, C. A. Walls, T. Parish, J. L. Holm, P. A. Morris, S. S. Walls and F. I. Doyle. *J. Zool., Lond.* 244, 7-21 (1998).
[19] P. W. W. Lurz, P. J. Garson and S. P. Rushton. *For. Ecol Manage.* 79, 79-90 (1995).
[20] J. Gurnell and H. Pepper. *Perspectives on the management of red and grey squirrels.* In: Jardine DC, editor. *Wildlife management in forests.* Edinburgh, Institute of Chartered Foresters; Edinburgh; 1988; pp 92-109.
[21] J. Gurnell and H. Pepper. *Mamm. Rev.* 23, 127-137 (1993).
[22] D. Parrott, R. Quy, K. Van Driel, P. W. W. Lurz, S. Rushton, J. Gurnell, N. Aebischer, J. Reynolds. *Review of Red Squirrel Conservation Activity in Northern England.* Natural England, Peterborough, (2009).
[23] P. W. W. Lurz, N. Geddes, A. J. Lloyd, M. D. F. Shirley, B. Burlton and S. P. Rushton. *Forestry* 76, 95-108.
[24] Red Squirrels Northern England. Conservation and strongholds. Accessed 9.12.2013. URL: http://www.rsne.org.uk/conservation-and-strongholds.
[25] Forestry Commission Scotland. Strongholds. Accessed 9.12.2013. URL: http://www.forestry.gov.uk/forestry/infd-7q3ft8.
[26] P. W. W. Lurz. *Ecology and Conservation of the Red Squirrel (Sciurus vulgaris L.) in Upland Conifer Plantations.* PhD Thesis, Newcastle University, Newcastle, UK (1995).
[27] P. W. W. Lurz. *Kidland Squirrel Monitoring.* Forestry Commission, Kielder Forest District, Bellingham, UK, (2013).
[28] P. W. W. Lurz, P. J. Garson and J. F. Ogilvie. *Forestry* 71, 67-71 (1998).
[29] P. W. W. Lurz, P. J. Garson and L. Wauters. *J. Zool., Lond.* 251, 167-178.
[30] E.-D. Schulze. *Science* 244, 776-783 (1989).
[31] S. Hagner. *Studia Forestalia Sueica* 33, 1-23.
[32] B. G. Hibberd. *Forestry Practice.* Forestry Commission Handbook No. 6. HMSO, London, (1991).

[33] J. D. Matthews. *Silvicultural Systems*. Oxford Science Publications, Oxford, UK (1989).
[34] S. J. Petty, I. J. Patterson, D. I. K. Anderson, B. Little and M. Davison. *For. Ecol. Manage* 79, 133-146 (1995).
[35] A. Broome, S. Hendry and A. Peace. *Forestry* 80, 17-28 (2007).
[36] E. Pulliainen and L. H. J. Lajunen. *Can. J. For. Res*. 14, 214-217 (1983).
[37] S. J. Petty, P. W. W. Lurz and S. P. Rushton. *Biol. Cons*. 111, 105-114 (2003).
[38] J. P. LaRose, A. L. Meredith, D. J. Everest, C. Fiegna, C. J. McInnes, D. J. Shaw and E. M. Milne. *Vet. Rec*. 167, 297-302.
[39] C. M. Shuttleworh and J. Gurnell. *The management of coastal sand dune woodland for red* squirrels (*Sciurus vulgaris L.).* In: Houston J, Edmondson ASE and Rooney PJ, editors. Coastal Dune Management: *Shared Experience of European Conservation Practice*. Liverpool University Press; Liverpool; 2001; pp 117-127.
[40] L. A. Wauters and A. A. Dhondt. *Anim. Behav*. 43, 297-311 (1992).
[41] P. W. W. Lurz, P. J. Garson and L. Wauters. *Anim. Behav*. 54, 427-435 (1997).
[42] S. Bosch and P. W. W. Lurz. *The process of drey construction in red squirrels* in press. Hystrix (2014).
[43] J. Bryce, S. Cartmel and C. P. Quine. *FC Inf. Note* 76 (2005).
[44] J. Bryce, P. J. Johnson and D. W. MacDonald. *J. Appl. Ecol*. 39, 875-887.
[45] S. Cartmel. A study of red and grey squirrels in Clocaenog Forest, North Wales. CCW, Wales (1999).
[46] S. P. Rushton, J. Gurnell, P. W. W. Lurz and M. D. F. Shirley. *Spatially explicit modelling approach to identify potential areas for conserving red squirrels in mainland Britain*. JNCC, Peterborough, UK (1999).
[47] C. Shuttleworth and T. Donoyou. *Red Squirrel Population Monitoring Clocaenog*. Forestry Commission Wales (2012).
[48] C. M. Shuttleworth. *Quart. J. For*. 98, 113-119 (2004).
[49] J. P. Duff, R. Higgins and S. Farrelly. *Vet. Rec*. 160, 384 (2007).
[50] D. J. Everest, S. S. Grierson, M. F. Stidworthy and C. Shuttleworth. *Vet. Rec*. 165, 482 (2009).
[51] D. J. Everest, C. M. Shuttleworth, S. S. Grierson, J. P. Duff, N. Jackson, P. Litherland, R. E. Kenward and M. F. Stidworthy. *Vet. Rec*. 171, 176 (2012).
[52] D. J. Everest, H. Butler, T. Blackett, V. R. Simpson and C. M. Shuttleworth. *Vet. Rec*. 173, 199-200 (2013).

[53] A. White and P. W. W. Lurz. *A modelling assessment of control strategies to prevent/ reduce Squirrelpox spread.* Scottish Natural Heritage Report, Edinburgh, UK (2014).
[54] Anonymous. *FC Practice Note* 102, 1-16 (2012).
[55] C. M. Shuttleworth, P. W. W. Lurz, N. Geddes and J. Browne. *For. Ecol. Manage.* 279, 167-175.
[56] A. Brown and J. Webber. *FC Research Note* 002, 1-8 (2008).
[57] Forestry Commission. URL: http://www.forestry.gov.uk/forestry/INFD-8EJKP4. Accessed 10.1.2014.
[58] D. Ray. *FC Research Note* 101, 1-8 (2008).

In: Forest Ecosystems
Editor: Noel C. Roberts
ISBN: 978-1-63117-815-3

Chapter 4

IMPACT OF WOOD STRUCTURE AND SURFACE MORPHOLOGY ON HEAT AND MASS TRANSFER RESISTANCE AT NORWAY SPRUCE

*A. Straže**
University of Ljubljana, Biotechnical Faculty, Ljubljana, Slovenia

ABSTRACT

The heat- and water vapor transfer in wood and the closely linked wood structural properties have been studied extensively using steady state conditions. The experimentation on Norway spruce confirmed significant impact of surface morphology on established boundary conditions and there from on changed heat in mass transfer. The real bearing wood surface profiles actually often follow Abbot curve with a signed increase of bearing surface towards the inner wood tissue. This leads to the ascertaining of the inverse proportional relation between the contact thermal- and surface mass transfer resistance and the contact surface of wood. The results confirm the existence of contact thermal- and surface mass resistance when using the hot plate method or the steady state diffusion Cup method. Findings support discussion on the credibility and careful use of steady state experimentation on heat- and mass transfer at Norway spruce wood.

---

* Email: ales.straze@bf.uni-lj.si.

Keywords: Wood, surface morphology, heat and mass transfer, surface resistance, thermal conductivity, water diffusion

STRUCTURAL PROPERTIES OF NORWAY SPRUCE WOOD

The anatomical structure of Norway spruce wood (*Picea abies* Karst.) consists only of two anatomical cell types: tracheids and parenchyma cells. Tracheids are the main structural component of Norway spruce wood and their percentage in the wood ranges between 94.5 in 96.5% (Wagenführ 2007; Timell 1986). The remaining percentage is composed of parenchyma cells which form the pith rays, the axial parenchyma and resin canals (Plomion et al., 2001) (Figure 1).

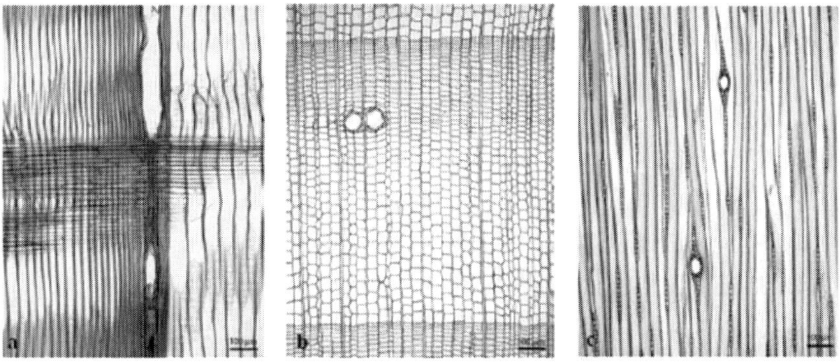

Figure 1. The typical oriented light microscopy slices of Norway spruce wood: a) RL-plane, b) RT-plane, c) TL-plane.

Geometrical Cellular Model of Norway Spruce Wood

Building of the geometrical model of real wood, irrespective of the relative simplicity anatomy of Norway spruce wood, often has some limitation. Most commonly the model presents just the average structure, usually established by microscopic observation. We made at the Norway spruce wood for this purpose light microscopy slice at random locations of real tested specimens (Figure 1) wherefrom, using image analysis software wood cells dimensions were determined (Bučar and Straže 2008). Figure 2 shows a geometric model built from the average dimensions of the early and late wood cells of the Norway spruce samples used. Within the examined specimens, the early wood represented 85.3% of the volume, and late wood 14.7% of the porous structure. Cell dimensions were calculated from the average of 100 measurements on a microscopic preparation, with a low coefficient of variation (CV ≤ 10%), and they were comparable to the results of other research (Timell 1986; Wagenführ 2007). The determined geometrical model (Figure 2c), by the propagation of geometrical wood cells of early- and late wood, was afterwards used to build a full-scale geometrical model (Figure 2b) of real wood specimens (Figure 2a).

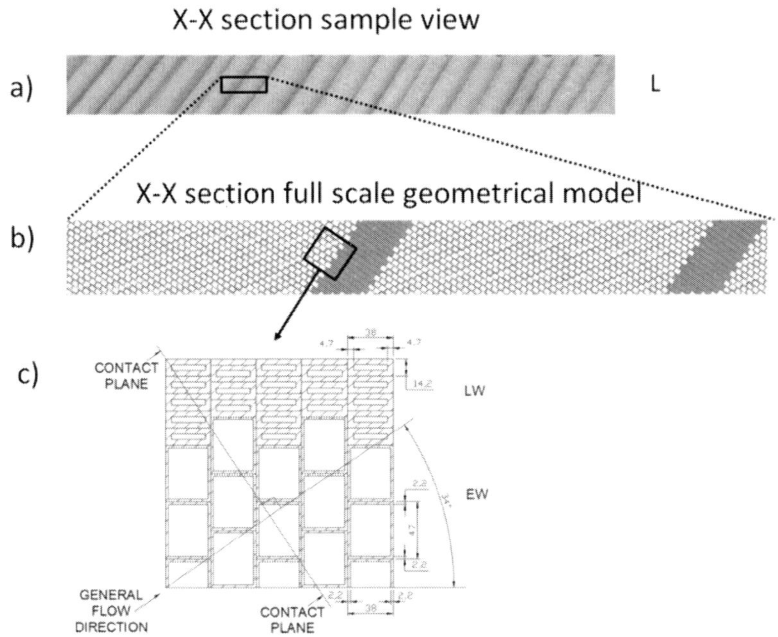

Figure 2. The Norway spruce wood samples' cross section (a) used for building of geometrical cellular model (b) made from average earlywood- and latewood cells (c) (Bučar and Straže 2008).

Full scale wood specimens, having the identical cell cross-section orientation of the geometrical model, were used for heat and mass transfer experimentation in axial- and transversal direction of the wood fibers. The experiments were made in various thicknesses of the specimen; to analyze the impact of internal and external heat- and mass transfer resistance, where the latter, according to studies (Choong and Skaar 1972, 1969; Wolf and

Schneider 1998; Ciavarella 2004) depends also on specimen surface morphology (Figure 3).

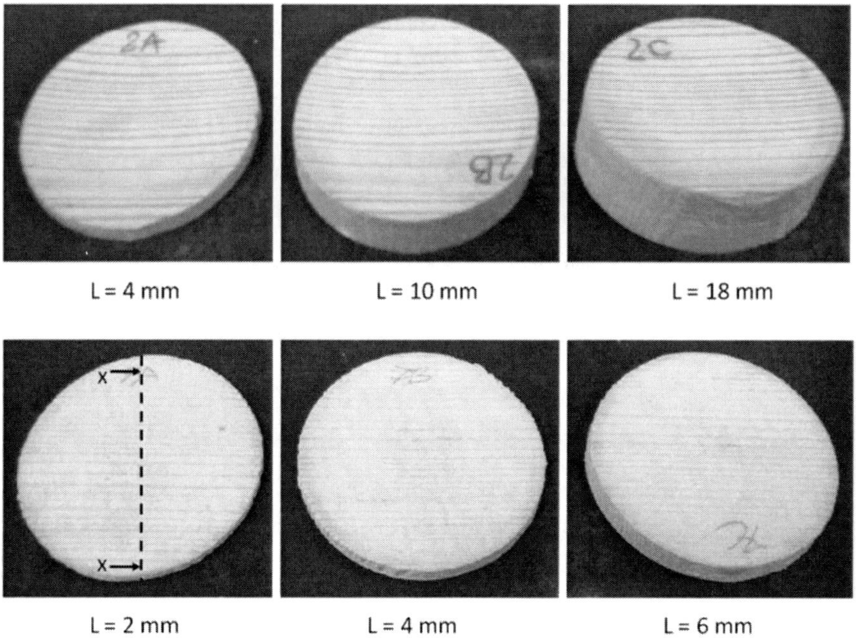

Figure 3. The parallel samples of Norway spruce wood, having various thickness, for heat- and mass transfer testing in longitudinal (above) and transversal direction (bottom) (Bučar and Straže 2008; Straže and Gorišek 2011).

Physical Characterization of Wood Surface

The production of wood products generally needs a series of transformation or modification processes so that each one enables any reduction of the size by the cutting of the wood. The cutting process, also ablation or melting is determined by the machining parameters and surface roughness (Aguilera and Martin 2001). The wood surface directly concerns the further use of wood. If the wood surface roughness is known, its quality for the specific function can be determined.

Surface Characteristics of Ideal Norway Spruce Wood

The ideal contact surface, which lies in the theoretical wood plane, can be determined by cell model (Figure 2c). It is assumed that the majority of heat flow runs through the cell structure whereby the contribution of convection through lumens to the total flow is negligibly small. Generally, the ideal contact surface of the wood is decreased by at least the quantity of empty spaces in the examined plane. According to the established model, as a case study (Figure 2c), the ideal contact surface in the direction of wood fibers was decreased to 19.8% of the nominal surface area in the earlywood, and to 74.3% of the nominal surface area in the latewood. Taking into account the proportions of both categories of wood, the ideal contact surface in the direction of wood fibers, therefore, represents only 29% of the nominal surface area of the specimen. Transversely to wood fibers, in the contact plane under the angle of 56° to the annual ring, the share of cell structure was 26% in the early wood and 77% in the late wood, while the rest was occupied by cell lumens. If – in this case as well – the proportion of both categories of wood is taken into account, the ideal contact surface area in this plane occupied 33% of the total porous surface area of the tissue. It turned out that the test specimens had similar ideal contact surfaces in the longitudinal and transverse direction of the wood fibers.

The Impact of Cutting Process and Wood Structure on the Wood Surface Characteristics

The real, mechanically treated, wood surface combines the elements of an ideal contact surface with the added effect of mechanical treatment (Gurau et al., 2005; Milanez et al., 2003). The relief of an ideal contact surface is defined in detail with the geometry of the wood cells and also with the morphology of the cell walls well. The collateral result of many real mechanical treatments of wood is often kneading and collapsing of the cell structures, the spring of inter- and intracellular fractures and also pulling the wood fibers (Jokerst and Stewart 1976; Murmanis et al., 1983). The morphology of the real wood surface is therefore significantly different compared to the ideal cross section of wood tissue in a particular anatomical plane. The fact is that the real contact- or exposed hygroscopic surface of the porous wood is significantly different from the geometrically ideal contact surface, where the latter is certainly smaller compared to the nominal surface of the wood specimen.

Figure 4 presents the typical surface appearance of Norway spruce wood after precise saw-blade cutting in RL- and RT-plane, found in our tested material, having raised wood tissue with poor connection to the underlying fibrous basis.

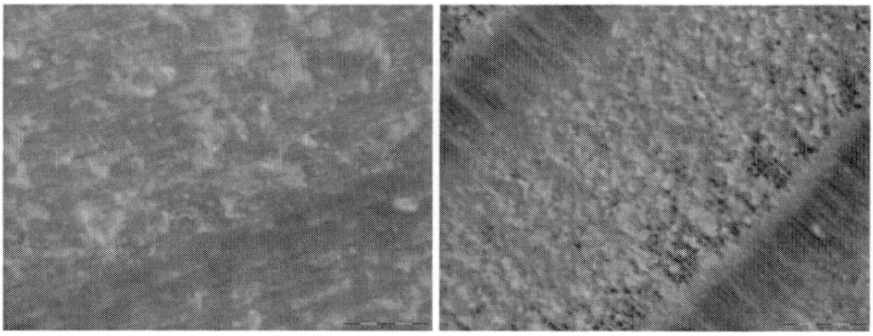

Figure 4. The real surface appearance of Norway spruce wood after precise saw-blade cutting in RL- (left) and RT-plane (right) (dash = 100 μm) (Bučar and Straže 2008; Straže and Gorišek 2011).

Recently, regarding quality of contact wood surface also contrary assertions were noticed. Namely, several studies report successful use of laser technology to remove microscopic amounts in order to clean or prepare the surface for microscopical observations or for coatings and other potential applications by controlling of energy input (Panzner et al., 1998; Seltman 1995). The laser technology permits a precise control of the energy input, which can be varied by changing the wavelength of the beam, pulse duration, repetition rate and intensity so that the anatomy of the wood surface can be modified within the range from laying bare the cellular structure to sealing the surface by melting and solidification. This technology actually uses melting of wood, in the range from 140 to 200 °C, where the glass transition point of lignin and cellulose exist respectively (Sandermann and Augustin 1963a, 1963b). The surface structure obtained from this type of ablation can be even similar to that of microtome cut for SEM investigation.

Morphology of the Real Contact Surface of Norway Spruce Wood

Longitudinal and transverse real contact surfaces of the tested specimens (Figure 3 and 4) are presented with a typical line profile of surface roughness and its vertical frequency distribution (Figure 5 and 6). Profiles deviated significantly in both wood directions from the properties of an ideal geometrical model. The vertical variation of the measured profiles significantly exceeded the limits of the geometrical model.

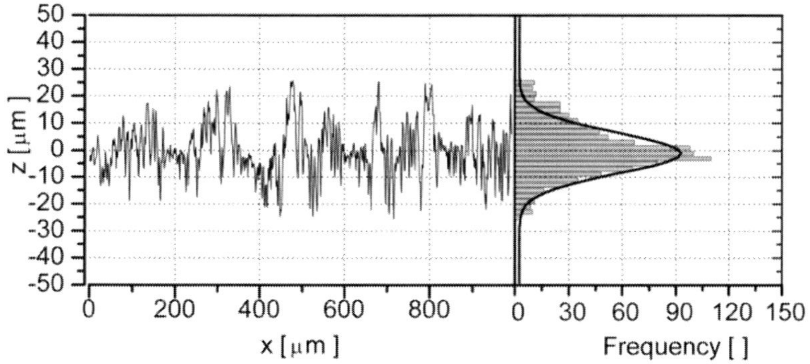

Figure 5. Line profile of surface roughness in longitudinal direction of Norway spruce wood (RL-plane) with vertical frequency distribution and comparison with Gaussian (Bučar and Straže 2008).

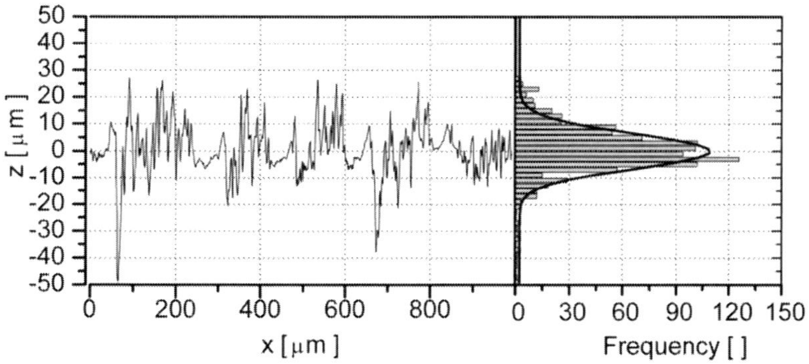

Figure 6. Line profile of surface roughness in transverse direction of Norway spruce wood (RT-plane) with vertical frequency distribution and comparison with Gaussian (Bučar and Straže 2008).

We assume that short and smoother intervals of the line surface profiles represent latewood zones, whereas intervals with higher peaks represent early wood zones on the surface of the test specimens. The vertical frequency distribution of the line profiles has been just partly normal, with standard skewness and kurtosis out of the range of 2. We determined the prevailing part of the vertical frequency distribution of both longitudinal and transverse surfaces below the mean value of profiles, which confirms the asymmetry of vertical profiles of the surface roughness in the direction of the inner wood structure.

The cumulative bearing length of a typical line profile of both longitudinal and transverse surface roughness is presented by the Abbot curve in Figure 7. The maximal contact surface, equal to the nominal one, is expected at the lowest values of the line profiles. The sigmoid cumulative bearing length decreases in the vertical direction, and reaches the zero value at the upper boundary of the line profiles. No difference in the Abbot curves of both transverse and longitudinal wood surfaces is confirmed.

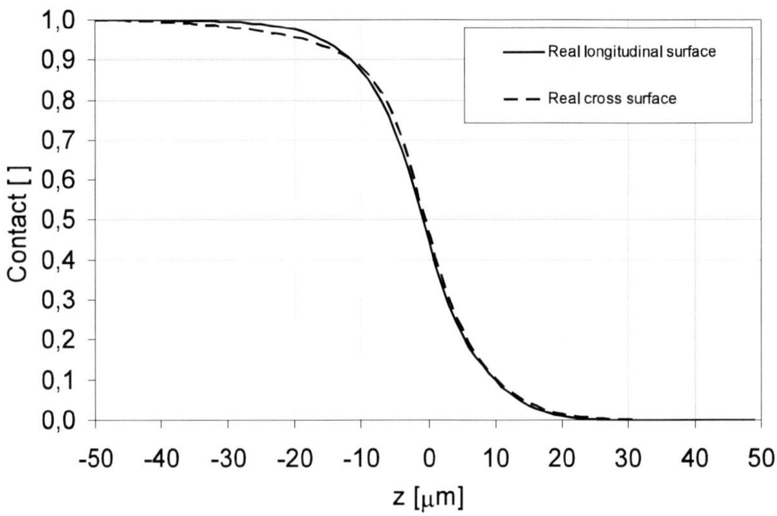

Figure 7. Abbot curve with average cumulative bearing capacity of real Norway spruce wood surface in longitudinal (——) and transverse direction of wood (- - -) (Bučar and Straže 2008).

HEAT AND MASS TRANSFER AT WOOD IN STEADY STATE CONDITIONS

Steady State Heat- and Mass Transfer Experimentation

Several studies have dealt with the thermal- and bound water conductivity of wood in the past, using often identical methods to model both types of transport (Peralta and Skaar 1993; Suleiman et al., 1999; Khattabi and Steinhagen 1993), where also heat flow was defined by the laws of diffusion. The majority of data about the thermal conductivity of wood was obtained by applying the conventional stationary experimental method using a hot plate (Wangaard 1969; Kolmann and Malmquist 1956; Suleiman et al., 1999; Kamke and Zylkowski 1989), whereas the bound water transport has been studied extensively using Fick's first law of diffusion with data from steady-state experiments, traditionally called the cup method (Comstock 1963; Choong 1965; Olek 2003).

The inverse value of total resistance (R_t) against heat- and mass transfer were at both methods, also due to simplicity of mathematical solution, initially ascribed to thermal conductivity (λ) and to bound water diffusivity of wood (D). Further research disclosed later on various supplemental influencing porous-material factors, such as the presence of thermal diffusion of gases and material thermal dilatation due to long lasting heat flow at hot-plate method (Suleiman et al., 1999; Siau and Babiak 1983; Siau et al., 1986). Similarly the total mass transfer resistance at diffusion cup was often significantly influenced by reducing vapor concentration potential due to the distance of a wood sample from the salt (concentration) solution in the cup and also by strong dependence of water-vapor diffusion in wood on the vapor concentration, i.e., the wood moisture content (Choong 1965; Olek 2003; Stamm 1960; Straže 2010).

External Heat Transfer Resistance at Steady State Experimentation on Norway Spruce Wood

The experiments on determination of thermal conductivity of wood revealed the existence of additional surface / contact thermal resistance when using hot-plate method. Contact thermal resistances result from the morphological and anatomical properties of surfaces in contact, and these properties influenced the ratio of real to the nominal contact surface area (Bučar and Straže 2008; Wolf and Schneider 1998; Ciavarella 2004; Bučar et al., 2005). Consideration of the contact thermal resistance when using the hot plate method was particularly necessary when measuring thin wood specimens with rough surfaces and in the direction of wood fibers where the thermal resistance had the lowest values (Bučar and Straže 2008).

The simultaneous influences of the deformability of the material, local hardness and geometrical properties of surfaces had a very important role as well. Increasing normal loads and the elasto-plastic deformability of participating structures decrease the contact thermal resistance (Jeng et al.,

2006; Buonanno et al., 2003). Increasing contact surface between the wood and the smoother, less deformable plates of hot plate devices is expected when moving towards the inner wood tissue, according to the material / wood bearing capacity. Considering the inverse proportionality of the contact thermal resistance and contact surface (Ciavarella 2004; Yeh et al., 2003; Wolf and Schneider 1998), the similar decrease of contact thermal resistance on both transverse and longitudinal surfaces of Norway spruce wood was confirmed when moving the contact plane towards the inner wood tissue (Bučar et al., 2005; Bučar and Straže 2008). The minimal contact thermal resistance of wood is therefore theoretically reachable at the maximal bearing capacity of the wood surface, using sufficiently high contact pressure, when it depends only on the morphological characteristics of the wood cell walls.

External Mass Transfer Resistance at Steady State Experimentation on Norway Spruce Wood

The credibility of the cup method for the determination of the diffusion coefficient (*D*) was similar to the hot-plate method often questioned since it has been experimentally proved that the wood surface moisture content differs significantly from the equilibrium moisture content (*EMC*) (Shmulsky et al., 2002; Comstock 1963). Therefore, correction of the relative humidity (*RH*) at the surface of the tested specimen was often calculated (Siau 1984). Additionally, the physical interpretation of the Fick's first law equation means that moisture content (*MC*) of wood surface has to be immediately equal to the *EMC*. The consequence of this assumption is neglecting the so called external or surface resistance of moisture transfer (R_s), which is questionable for small size wood specimens, especially for the longitudinal moisture transport determination (Siau and Avramidis 1996). The water mass flow at the diffusion cup method can therefore be determined in the word specimen and also at the surface boundary layer (Eq. 1).

$$\dot{m} = D \cdot A \cdot \frac{\Delta c_1}{L} = S \cdot A \cdot \Delta c_2 \tag{1}$$

where *A* and L represent cross section and the thickness of the specimen, L = thickness of the specimen [m], D = diffusion coefficient [m^2/s], S = surface emission coefficient [m/s], Δc_1 and Δc_2 are moisture concentration difference between two surfaces of the specimen and between the surface of the specimen and the surroundings, respectively.

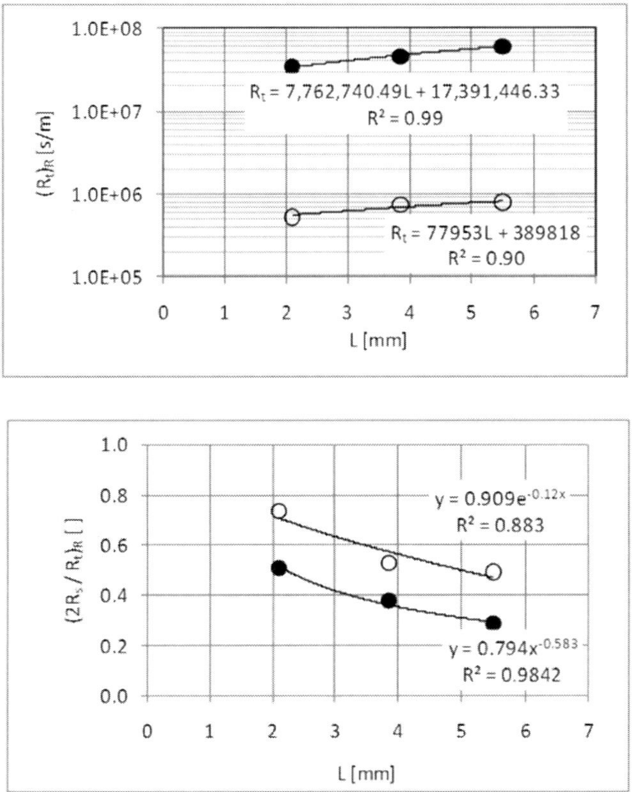

Figure 8. The dependence of the total mass transfer resistance (R_t) (left) and share of external mass transfer resistance ($2R_s/R_t$) (right) on the thickness of the specimen (L) in the perpendicular direction (● - low average *MC*, o - high average *MC*) (Straže and Gorišek 2011).

Similar to electrical circuits, resistances in moisture transport circuits, as well as at heat transport, can also be simply added up in accordance with the way they are connected. In the case of diffusion cup method, there are in addition to the basic internal diffusion resistance of the specimen ($R_i = L/(D \times A)$) – also two external, surface resistances ($R_s = 1/(S \times A)$), which can either be equal or different. The total mass transfer resistance (R_t) can therefore be written as a function of the thickness of the specimen having constant parameter, which represent two external resistances ($2/S$):

$$R_t \cdot A = \left(2 \cdot R_s + R_i\right) \cdot A = 2\frac{1}{S} + \frac{L}{D} \tag{2}$$

Recent research using the above methodology at steady state diffusion experiments on Norway spruce wood in perpendicular and longitudinal wood direction, varying sample thicknesses and moisture content, confirmed significant influence of external resistance to the total water transfer (Straže and Gorišek 2011).

The share of R_s was the highest at the lowest thickness of specimens, and exponentially decreased at higher thicknesses. Oppositely to the total mass transfer resistance, external mass transfer resistance decreased with reduction of wood MC (Figure 8 and 9).

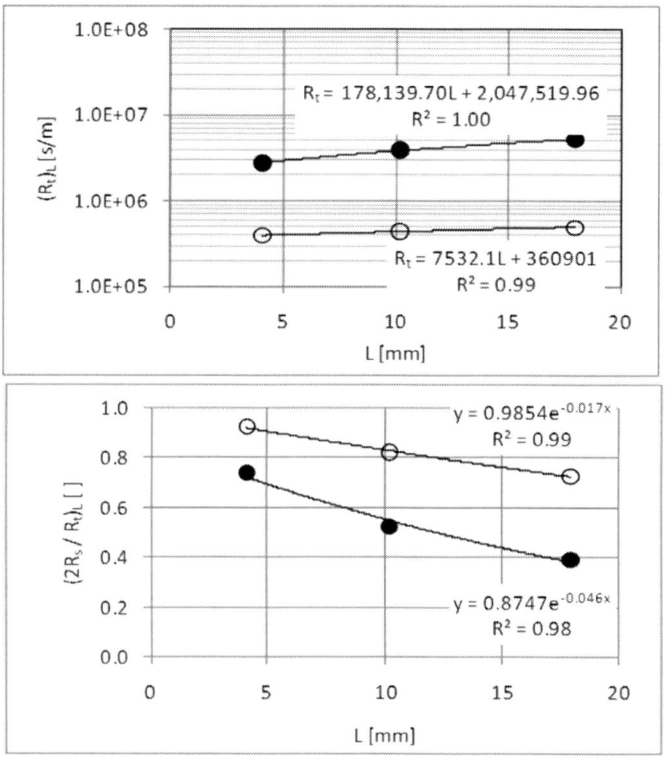

Figure 9. The dependence of the total mass transfer resistance (R_t) (left) and share of external mass transfer resistance ($2R_s/R_t$) (right) on the thickness of the specimen (L) in the longitudinal direction (● - low average *MC*, o - high average *MC*) (Straže and Gorišek 2011).

Conclusion

Studies confirmed the existence and significance of contact thermal resistance and surface mass transfer resistance when using the hot plate- or diffusion cup method to determine the thermal- and mass transfer properties of wood. Inconsideration of the surface resistance when using these methods is particularly problematical at measuring of thin wood specimens with rough surfaces and in the direction of wood fibers where the thermal- and mass resistances have the lowest values. In such cases the resulting high share of contact thermal- and surface mass transfer resistances in the total resistance can cause significant measurement errors.

Both, the contact thermal- and surface mass transfer resistances of wood depend on the morphology of the surface. Contact thermal resistance at the contact of plane wood surfaces with morphologically less coarse materials depends only on the size of the contact surface area of the wood. The contact thermal resistance decreases with an increasing contact surface of the wood. Surface mass transfer resistance is in addition dependent also on the thermodynamic conditions of the gas-medium in the contact with the wood surface. Generally, the surface mass transfer resistance in case of wood–air contact decreases with the reduction of water vapor pressure, especially with increasing of the air the velocity at the wood surface boundary layer.

References

Aguilera A, Martin P (2001) Machining qualification of solid wood of *Fagus sylvatica* L. and *Picea excelsa* L.: cutting forces, power requirements and surface roughness. *Holz als Roh- und Werkstoff* 59:483-488.

Bučar B, Merhar M, Gospodarič B (2005) The problem of contact heat resistance in determining stationary monoaxial heat conductivity of transversely oriented wood tissue. *Les/Wood* 57 (3):52-56.

Bučar B, Straže A (2008) Determination of the thermal conductivity of wood by the hot plate method: The influence of morphological properties of fir wood (*Abies alba* Mill.) to the contact thermal resistance. *Holzforschung* 62:362-367.

Buonanno G, Giovinco G, Massarotti N (2003) An elasto-plastic model of thermal contact conductance between nominally flat surfaces in vacuum. *Int Comm Heat and Mass Transfer* 30 (7):921-930.

Choong ET (1965) Diffusion coefficients of softwoods by steady-state and theoretical methods. *Forest Products Journal* 15:21-27.

Choong ET, Skaar C (1969) Separating internal and external resistance to moisture removal in wood drying. *Wood Science* 1:200-202.

Choong ET, Skaar C (1972) Diffusivity and surface emissivity in wood drying. *Wood and Fiber* 4:80-86.

Ciavarella M (2004) The electrical/thermal conductance of rough surfaces – the Weirstrass-Archard multiscale model. *International Journal of Solids and Structures* 41:4107-4120.

Comstock GL (1963) Moisture diffusion coefficients in wood as calculated from adsorption, desorption, and steady state data. *Forest Products Journal* 13 (3):97-103.

Gurau L, Williams HM, Irle M (2005) Processing roughness of sanded wood surfaces. *Holz als Roh- und Werkstoff* 63:43-52.

Jeng YR, Chen JT, Cheng CY (2006) Thermal contact conductance of coated surfaces. *Wear* 260:159-167.

Jokerst RW, Stewart HA (1976) Knife- versus abrasive-planed wood: Quality of adhesive bonds. *Wood and Fiber Science* 8 (2):107-113.

Kamke FA, Zylkowski SC (1989) Effects of wood-based panel characteristics on thermal conductivity. *Forest Products Journal* 39:19-24.

Khattabi A, Steinhagen P (1993) Analysis of transient nonlinear heat conduction in wood using finite-difference solutions. *Holz als Roh- und Werkstoff* 51 (4):272-278.

Kolmann F, Malmquist L (1956) The coefficient of thermal conductivity of wood and wood based materials. *Holz als Roh- und Werkstoff* 14:201-204.

Milanez FH, Yovanovich MM, Culham JR (2003) Effect of surface asperity truncation on thermal contact conductance. *IEEE Transactions on Components and Packing Technologies* 26 (1):48-54.

Murmanis L, River BH, Stewart HA (1983) Microscopy of abrasive-planed and knife-planed surfaces in wood-adhesive bonds. *Wood and Fiber Science* 15 (2):102-115.

Olek W (2003) Analysis of the cup method application for determination of the bound water diffusion coefficient in wood. *Folia Forestalia Polonica* 34 (2):15-25.

Panzner M, Wiedemann G, Henneberg K, Fischer R, Wiitke T, Dietsch R (1998) Experimental investigation of the laser ablation process on wood surfaces. *Applied Surface Science* 129:787-792.

Peralta PN, Skaar C (1993) Experiments on steady-state nonisothermal moisture movement in wood. *Wood and Fiber Science* 25 (2):124-135.

Plomion CH, Leprovost G, Stokes A (2001) Wood formation in trees. *Plant Physiology* 127 (4):1513-1523.

Sandermann W, Augustin H (1963a) Chemische Untersuchungen über die thermische Zersetzung von Holz: Erste Mitteilung - Stand der Forschung. *Holz als Roh- und Werkstoff* 21:256-268.

Sandermann W, Augustin H (1963b) Chemische Untersuchungen über die thermische Zersetzung von Holz: Zweite Mitteilung - Untersuchungen mit Hilfe der Differential-Thermo-Analyse. *Holz als Roh- und Werkstoff* 21:305-319.

Seltman J (1995) Freilagen der Holzstruktur durch UV Bestrahlung. *Holz als Roh- und Werkstoff* 53:225.

Shmulsky R, Kadir K, Erickson R (2002) Effect of air velocity on surface EMC in the drying of red oak lumber. *Forest Products Journal* 52 (1):78-80.

Siau JF (1984) Transport processes in wood. Springer-Verlag, Berlin.

Siau JF, Avramidis S (1996) The surface emission coefficient for wood. *Wood and Fiber Science* 28:178-185.

Siau JF, Babiak M (1983) Experiments on nonisothermal moisture movement in wood. *Wood and Fiber Science* 15 (1):40-46.

Siau JF, Bao F, Avramidis S (1986) Experiments on nonisothermal diffusion of moisture in wood. *Wood and Fiber Science* 18 (1):84-89.

Stamm AJ (1960) Combined bound-water and water-vapour diffusion into sitka spruce. *Forest Products Journal* 10:644-648.

Steinhagen P (1977) Thermal properties of wood, green or dry, from -40° to +100 °C: A literature review. USDA Forest Service, Madison.

Straže A (2010) Impact of external and internal mass transfer resistance on kinetics of convective wood drying. Doctoral Dissertation, University of Ljubljana, Ljubljana.

Straže A, Gorišek Ž Analysis of external and internal mass transfer resistance at steady state diffusion experiments on small clear wood specimens. In: Hofstetter K (Ed) Mixed numerical and experimental methods applied to the mechanical characterization bio-based materials: COST Action FP0802 thematic workshop, Vila Real, April 27-28 2011. CITAB, pp 23-24.

Suleiman BM, Larfeldt J, Leckner B, Gustavsson M (1999) Thermal conductivity and diffusivity of wood. *Wood Science and Technology* 33 (6):465-473.

Timell TE (1986) Compression wood in gymnosperms, Vol 1. Springer, Berlin.

Wagenführ R (2007) Holzatlas. Hanser, Berlin.

Wangaard FF (1969) Heat transmissivity of Southern pine wood, plywood, fiberboard and particleboard. *Wood Science* 2 (1):54-60.

Wolf EG, Schneider DA (1998) Prediction of thermal contact resistance between polished surfaces. *International Journal of Heat and Mass Transfer* 41:3469-3482.

Yeh CL, Chen YF, Wen CY, Li KT (2003) Measurement of the thermal contact resistance of aluminum honeycombs. *Experimental Thermal and Fluid Science* 27:271-281.

In: Forest Ecosystems
Editor: Noel C. Roberts
ISBN: 978-1-63117-815-3

Chapter 5

QUANTIFYING THE EFFECT OF PLANT DIVERSITY AND MATURITY ON SOIL EROSION ON DIFFERENT SPATIAL SCALES IN YUNNAN PROVINCE OF CHINA

***Zhenhong Wang*[1,*] *and Chuanqun Duan*[2]**
[1]College of Life Science, Guizhou University, Guiyang City, China
[2]College of Life Science, Yunnan University, Kunming City, China

ABSTRACT

Elucidating the relationship between plant diversity and soil erosion on different scales can give insights into the interactions between botanical processes and the earth's surface, and is crucial for the establishment of theories on biodiversity and ecosystem functions. In Yunnan province, China, we studied local ecosystems that showed gradual changes in plant diversity and ecosystem maturity, watershed ecosystems that only presented gradual changes in ecosystem maturity, and 11 floral subregions, representing regional ecosystems, in which plant diversity was assessed based on the recorded statistics of 13,850 seed plants. Runoff plots and weir method were used to determine soil erosion in the local and watershed ecosystems from 2000 to 2006. Soil erosion of the regional ecosystems was determined by superposing about 10,000 patches of soil-erosion maps, with predetermined erosion grades,

* Corresponding author's Email address: w_zhenhong@126.com.

on the 12 floral subregions based on geographical information system (GIS). Surface runoff and soil erosion decreased with increasing plant diversity and ecosystem maturity in local ecosystems, but showed obvious differences under the conditions of similar plant diversity and different levels of ecosystem maturity in the watershed ecosystems. Soil erosion declined with increasing diversity in regional ecosystems. Soil erosion control was gradually enhanced with increasing diversity and ecosystem maturity and the function was mainly dependent on ecosystem maturity. Increasing diversity leads to higher plant cover and density in local ecosystems, thereby promoting erosion control. Covering, diffusion and density effects explain the enhanced process. Ecosystem maturity played a greater role in soil erosion control than plant diversity, which presented a weaker effect on the function. An enhanced function of soil erosion control with increasing diversity predicts an increase in ecosystem productivity. This study provides an integrated understanding of the relationships between diversity and ecosystem functions on different scales.

Keywords: Biodiversity; ecosystem function; soil conservation; watershed; floral subregion; surface runoff; soil erosion; stability

INTRODUCTION

Over the past 30 years, using theoretical approaches, field observations, and species-assemblage experiments, ecologists have strived to reveal the relationship between biodiversity and ecosystem functions, mainly productivity, sustainability, and stability. However, the mechanisms underlying the relationships between plant diversity and productivity, as well as other ecosystem functions, are still under debate (Schulze and Mooney 1993; Naeem et al. 1994; Tilman et al. 1996; Hector et al. 2001; Loreau et al. 2001; Szwagrzyk and Gazda 2007). Some ecologists have suggested that increased productivity is caused by ecosystem maturity, and that plant diversity does not have any effect on ecosystem function (Huston et al. 2000; Guo et al. 2006; Adler et al. 2011), bringing the study of plant diversity and ecosystem functions to a stalemate.

Nevertheless, the relationship between plant diversity and soil conservation, one of the 17 most important ecosystem functions influencing ecosystem sustainability, has never been systematically studied (Tilman et al. 1996; Costanza et al. 1997; Loreau et al. 2001). Ecologists still do not know if plant diversity can enhance soil conservation, and if it does, how it affects the ecosystem function (Spehn et al. 2000; Tilman 2000; Loreau et al. 2001). Soil conservation is defined as the prevention of soil loss via various removal processes from the perspective of ecosystem function (Costanza et al. 1997). Impairment of primary productivity generally occurs with soil erosion within an ecosystem, and soil erosion from local to global scales has led to significant decreases in primary productivity by causing the loss of soil nutrients (William et al. 1983; Lal 1988; Morgan 1996). We suggest that if plant diversity enhances soil conservation on different scales, and as a result, the ecosystem guarantees a nutritional supply for biomass production, this can be considered indirect proof of a positive relationship between diversity and productivity, thus breaking the stalemate. On the other hand, few studies have examined the relationship between plant diversity and ecosystem functions on a large scale, such as watershed or regional scales (Loreau et al. 2001; Szwagrzyk and Gazda 2007). Ecosystem processes differ with scale, and it is only when ecologists have fully understood the plant diversity- ecosystem function relationships on different scales that a theoretical framework for this relationship can be completely established.

Here we investigated plant diversity and soil erosion on local and watershed scales, and performed an indirect evaluation of soil erosion on a regional scale. We addressed the following questions: (1) Can increasing plant diversity enhance control of soil erosion on a local scale? (2) If so, how does plant diversity regulate function? (3) Which has the greater effect: plant diversity or ecosystem maturity? (4) How does the relationship between plant diversity and soil erosion behave on watershed and regional scales?

MATERIALS AND METHODS

Hypothesis

To answer question (1), we need to establish local plots that show gradual changes in plant diversity but are identical in other factors influencing soil erosion, and directly observe surface runoff, sediment and total P loss in the surface runoff. If these indexes are negatively related to plant diversity, we can conclude that soil conservation depends on plant diversity.

If plant cover and density are dependent on plant diversity in these plots, we can indirectly infer that soil conservation is positively related to plant diversity. Furthermore, we can analyze how plant diversity influences soil conservation (question 2), because plant cover and density have a direct positive effect on soil erosion control via kinetic-energy reduction, rainfall interception and soil strength (Lin et al. 1997; Calder 2001).

If we can establish a small watershed consisting of two sections, the upper and lower reaches, surface runoff and sediment from the two sections can be precisely measured: soil erosion factors are identical in these two sections, and plant diversity does not appear to be significantly different; however, ecosystem maturity differs. If the generation of surface runoff and sediment is not obviously different between the two sections, then plant diversity has a greater effect on soil conservation than ecosystem maturity. Conversely, maturity plays a greater role (questions 3 and 4) if surface runoff and sediment differ between the two sections. Note that if both plant diversity and ecosystem maturity differed between the two sections, we could not judge which plays a greater role.

Using data accumulated in long-term studies on plant species distribution in different regional ecosystems, plant diversity can be assessed using the statistics of the number of each species in these ecosystems. If data on soil erosion in these corresponding regions can be also collected, we can use geographical information system (GIS) techniques to superpose the soil-erosion data on the regions in which plant diversity has been assessed. The relationship between plant diversity and soil erosion can then be quantified on a regional scale (question 4).

Local Scale

Vegetation Survey

According to the diversity-age relationship, plant diversity and ecosystem maturity are predicted to show a gradual change along a secondary successional sere of a plant community (Ricklefs 2008). We therefore selected five secondary successional stages of semi-humid evergreen broadleaf forest at a local Muding site (101°28′18″E, 25°24′09″N, 1950–2015 m, Figure 1) as subjects in the Yunnan province of China. The site is 210 km away from Kunming.

Rainfall, which usually occurs between May and August, averages 880 mm per year. Average annual temperature is about 14.7°C. Purplish soils are the weathering product of purple sandshale.

Regionally, this soil type covers about 80% of floral subregions D and F in Figure 1, and the semi-humid evergreen broadleaf forest is mainly distributed in these regions.

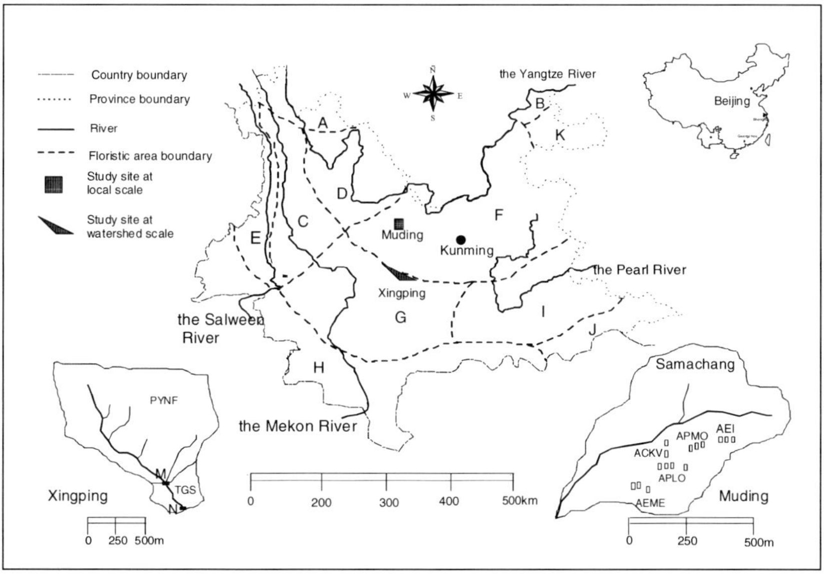

Figure 1. Location of local site, watershed, and region. The map shows the floral region of Yunnan province, China. Floral subregions are: (A) Kang-Tibet; (B) big-small Liangshang; (C) western Yunnan gorge; (D) Jingshajiang river; (E) Eastern-Himalaya; (F) Yunnan plateau; (G) Middle Liangcan-Honghe river; (H) Yunnan-Burma-Laos border; (I) eastern-south Yunan; (J) Yunnan-Vietnam border; (K) eastern Yunnan. Muding and Xingping in the bottom right- and left-hand corners of the figure represent the study sites at local and watershed scales, respectively. In the Muding figure, small rectangles represent the locations of permanent plots in different succession stages where sediment was observed. In the Xingping figure, and TGS indicate the two reaches of the watershed, and M and N represent the locations of rectangular weirs.

The five secondary successional stages of the zonal forest at the local site were: AEI (Ass. *Elsholtzia fruticosa* + *Imperata cylindrical* (Linn.) Beauv*,* a shrub-herb community with no tree layer); APMO (Ass. *Pinus yunnanensis* + *Myrsine africana* + *Oplismenus compsitus, Pinus yunnanensis* needle-leaf forest); APLO (Ass. *Pinus yunnanensis* + *Lithocarpus confines* + *Oplismenus compsitus, Pinus yunnanensis* needle-broadleaf forest); AEME (Ass. *Eucalyptus smithii* + *Myrsine africana* + *Eupatorium adenophorum,* a man-made forest); ACKV (Ass. *Cyclobalanopsis glaucoides* + *Keteleeria evelyniana* + *Viola duelouxii,* a fragmented semi-humid evergreen broadleaf forest). Because the hillsides have been closed for afforestation since the late 1980s, there is no obvious human disturbance at this site (Jiang 1984; Tang et al. 2007; Tang 2010).

We established three 10 × 40 m plots for the survey of tree layers in representative stands of each of the five stages (sampling area 6000 m^2, n = 15). Each plot was selected to have essentially the same slope angle (17 ± 2°), slope position and slope orientation (small map in the bottom right-hand corner of Figure 1). Soil texture and organic matter in the soil at 0–30 cm depth also did not present any obvious differences in these plots (Xu 2002; Li 2008). In each plot, four equally spaced 4 × 4-m and 1 × 1-m subplots were established along the diagonal line of the plot to obtain data of shrub (sampling area: 960 m^2, n = 60) and herbaceous (sampling area: 60 m^2, n = 60) layers.

We recorded all tree, shrub, and herb species names, and determined the numbers of each individual species. Five sampling sites were set up evenly along the diagonal line of each plot and soil bulk density was determined by cutting rings at 20–30 cm depth from the soil surface (A1 layer). Using CI-110 crown density was measured at 20 points along the diagonal line of each plot at a height of 20 cm, and plant cover was calculated for each plot by leaf area index analysis software based on the image of each point (n = 20). Species richness (S), Shannon-Wiener index (H), Pielou index (J), and ecosystem maturity (R_d) were calculated using equations (1), (2), and (3), respectively.

$$H = -\sum pi \ln pi \quad (i = 1, 2, 3\ldots s) \tag{1}$$

$$J = H / \ln s \tag{2}$$

$$R_d = [\prod_{i=1}^{n} (V_i / IV_i)]^{1/n} \tag{3}$$

where p_i is the proportion of the number of individuals of species i out of the total number of individuals of all species in the survey plot or subplot; s is species richness; n is the number of variables; Vi is the value of the ith variable; IV_i is the ideal value of the ith variable, that is, the value of the corresponding variable in the reference ecosystem. The chosen reference ecosystem was a mature and well-conserved semi-humid evergreen broadleaf forest near the study site. We selected aboveground biomass, species richness, soil bulk density and community similarity as the variables to calculate ecosystem maturity (n = 4).

Table 1. Plant diversity pattern of five successional stages on the local scale

Successional stages		APMO	AEME	AEI	APLO	ACKV
Tree layer	Species richness***	10 ± 3 (a)	9 ± 3(a)	5 ± 2 (b)	24 ± 5(c)	36 ± 5 (d)
	Shannon-Wiener index***	0.39 ± 0.30 (a)	0.19 ± 0.03 (b)	0.12 ± 0.02(c)	1.41 ± 0.32(d)	2.11 ± 0.34(e)
	Pielou index ***	0.39 ± 0.18(a)	0.16 ± 0.05(b)	0.14±0.03(c)	0.62 ± 0.12(d)	0.73 ± 0.04(d)
Shrub layer	Species richness**	14 ±5 (a)	21 ± 6 (b)	21 ± 7 (b)	24 ± 9 (c)	19±6(b)
	Shannon-Wiener index**	1.38 ± 0.12(a)	1.68 ± 0.41(b)	1.67 ± 0.33(b)	1.78 ± 0.23(c)	1.88 ± 0.30(b)
	Pielou index **	0.75 ± 0.07(a)	0.77 ± 0.12 (a)	0.76 ± 0.12(a)	0.75 ± 0.08(a)	0.69 ± 0.09(b)
Herb layer	Species richness***	30 ± 7(a)	27 ± 8 (b)	46 ± 10 (c)	25 ± 9 (d)	23 ± 5 (d)
	Shannon-Wiener index**	1.96 ± 0.41(a)	1.86 ± 0.51(b)	2.25 ± 0.27(c)	1.75 ± 0.20(d)	1.65 ± 0.40(e)
	Pielou index	0.75 ± 0.12	0.67 ± 0.08	0.65 ± 0.08	0.63 ± 0.17	0.60 ± 0.08
Species richness in the three layers***		43 ± 5 (a)	52 ± 3 (b)	58 ± 5 (c)	68 ± 4 (d)	73 ± 6 (e)
Ecosystem maturity***		0.42 ± 0.12 (a)	0.45 ± 0.09 (a)	0.52 ± 0.16(b)	0.63 ± 0.18(c)	0.68 ± 0.22(d)

** and *** indicate significant differences among the succession stages at α levels of 0.01 and 0.001, respectively. Diversity indices in the three layers were expressed as μ ± s.d. (tree: $n = 15$; shrub: $n = 60$; herb: $n = 60$). Different lowercase letters in parentheses between any two stages in each layer represent significant LSD-t at least α = 0.05. Identical lowercase letters indicate lack of significance. Species richness in the three layers is the mean of the number of different tree, shrub, and herb species in the three plots.

These variables are the marks of ecosystem maturity that are linked to ecosystem performance (Tilman et al. 1996), stability (Chapin et al. 2000), soil environment improvement (Wu 2003), and community similarity, respectively, of each stage to the reference ecosystem. The aboveground biomass of each stage was calculated using allometric models based on a vegetation survey (Feng 1999). Species richness was the mean of the number of different plant species in the tree, shrub, and herb layers in three plots at each stage. The bulk density was the mean of 15 samples in three plots at each stage. The community similarity $C_j = j/(a+b-j)$, where j is the number of species common to a particular stage and the reference ecosystem, and *a* and *b* are the numbers of all species in the *i*th stage and the reference ecosystem, respectively.

Species richness, the Shannon-Wiener index, and the Pielou index were significantly different among the five successional stages although a few indexes showed only a small difference. The species richness in all three layers and ecosystem maturity showed an obvious gradual change among succession stages (Table 1). Thus, the local site meets the study conditions.

Soil Erosion Observation

Fifteen runoff plots (10 m × 40 m) were established around the vegetation survey plots and isolated with waterproof barriers (concrete walls) to prevent water exchange. The barriers were 0.3 m high and descended to a soil depth of 0.7 m.

The plots were oriented such that the major axis was parallel to the longitudinal slope (small map in the bottom right-hand corner of Figure 1). A runoff-collection trough at the bottom of the plots channeled flow to runoff pools with a capacity of 1.2 m^3. After each runoff event, we thoroughly agitated the runoff in the pools, took 500-ml samples and stored them at 4°C until analysis for sediment concentration and total P (determined within 24 h of sampling). SW40 limnimeters were installed at the top of the runoff pool to measure water level in the pool for the determination of runoff volume. Surface runoff exceeding the 1.2-m^3 pool capacity flowed from a triangular weir fixed at the top of the runoff pool and its volume was calculated using pool water level as recorded by the limnimeter (Wang 1997). Total P in the samples was determined colorimetrically following complete nitric/perchloric acid digestion of unfiltered samples by the Murphy and Riley method (Wang 1997). Sediment concentration was measured by drying duplicate aliquots at 105°C and weighing the remaining solids. Rainfall was measured with a DSJ-2 siphon hyetometer installed on open land among the plots, and corrected to measurements from three ordinary hyetometers installed in other open-land areas within the study site. In total, we conducted 420 observations from 2000–2003 of all stages.

Data Analysis

We calculated the mean, standard deviation, and coefficient of variation (standard deviation × 100/mean) of all observations of surface runoff, sediment and loss of total P in the runoff (n = 84) for each stage. Smaller standard deviation and coefficient of variation indicate a more stable ecosystem function (Tilman 2000). We analyzed the respective relationships between ecosystem maturity and each of the three erosion indexes and species richness (representing plant diversity), and the relationship between species richness and ecosystem maturity, respectively, and plant density and plant cover, using SPSS V17. Analysis of variance (ANOVA) was used to identify the statistical differences. Recognizing the high degree of variability, a confidence level of 0.95 (α = 0.05) was selected as an indicator of significance.

Watershed Scale

Vegetation Survey

We first selected 10 watersheds, each covered by PYNF (*P. yunanensis* forest) and TGS (a shrub community) in their upper and lower reaches, respectively, in Xingping (101°54`E, 23°55`N, Figure 1). These watersheds had the same landform, mid-mountain sloping land, and soil type. The weighted average slope angle of each watershed was calculated based on the data for the slope angle and area of all small patches in each watershed (the slope angle and area of all of these patches in each watershed was determined using GIS 9.7 on the landform map). A vegetation survey was conducted in three selected watersheds with similar weighted average slope angle. We tested the same vegetation parameters as for the local plots using sample transects (1 m × 120 m).

The vegetation parameters in the herb layer were determined in six 1-m^2 plots, distributed evenly along the sample transects. The vegetation survey ended when we increased sample transects in the respective PYNF and TGS and could only see the recorded species. Species richness (*S*), Shannon-Wiener index (*H*), Pielou index (*J*), and ecosystem maturity (*Rd*) were calculated using equations (1), (2), and (3). We obtained a watershed in which plant diversity was not significantly different between PYNF and TGS (PYNF: 11 transects, n = 11, 1320 m^2 and TGS: 6 transects, n = 6, 720 m^2), but ecosystem maturity was significantly different (Table 2). This watershed thus met the study conditions in the hypothesis, whereas the Muding site watershed had too many land-use types. The weighted average slope angle of PYNF and TGS was, respectively, 26° and 24°. The upper reach of the watershed was 1.90 km^2 and the lower reach was 0.58 km^2 (small map in bottom left-hand corner of Figure 1).

Table 2. Plant diversity in the two watershed reaches, PYNF and TGS Lowercase letters in parentheses and asterisks are as in Table 1 (PYNF: *n* = 11 and TGS: *n* = 6). Species richness in the two layers is the mean of the number of different woody and herb species at the transect scale

Watershed sections		PYNF	TGS
Woody plant layer	Species richness	21±8(a)	25±7(a)
	Shannon-Wiener index	2.52±0.54(a)	2.62±0.67(a)
	Pielou index	0.84±0.08(a)	0.88±0.04(a)
Herbaceous plant layer	Species richness	9±2(a)	13±6(a)
	Shannon-Wiener index**	1.45±0.45(a)	2.03±0.84(b)
	Pielou index	0.74±0.20(a)	0.79±0.21(a)
Species richness in the two layers		34 ± 9 (a)	37 ± 5 (a)
Ecosystem maturity***		0.70± 0.21(a)	0.25 ± 0.08 (b)

The altitude ranged from 1765 to 2010 m, and rainfall averaged 953 mm per year. The annual average temperature was about 17.6°C. The soil was red earth, with little human disturbance. Regionally, this soil is distributed in all floral regions (Figure 1).

Soil Erosion Observation

Rectangular weirs were built at the outlet of PYNF and at the outlet of TGS (M and N, respectively, on small map in bottom left-hand corner of Figure 1). The water levels at the weirs were recorded by automatic SW40 limnimeters and the rate of flow was calculated using equation (4).

$$Q = 0.018Bh^{3/2} \tag{4}$$

where Q is the rate of flow ($m^3\ s^{-1}$), B is the width of the weir (m), h is the depth of the water flowing through the weir (m), and 0.018 and 3/2 are coefficients.

We sampled 500 ml of runoff every 30 min during the runoff events, and kept at it 2°C. The sediment in the runoff was measured by the oven-drying method. We calculated the runoff volume and soil erosion based on the rate of flow and sediment concentration. The data at spot M illustrated the runoff volume and soil erosion in watershed PYNF, and the data at N showed runoff volume and soil erosion of PYNF + TGS (total watershed). The data at spot N minus that at spot M equaled runoff volume and soil erosion in TGS. Rainfall was measured by a DSJ-2 siphon hyetometer installed in the watershed. The accuracy of the siphon hyetometer was checked by three ordinary hyetometers installed on other open-land areas in the watershed. Because rainfall records did not present significant differences among hyetometers, we also ignored the influence of rainfall differences between PYNF and TGS on soil erosion. These observations were conducted from 2004–2006 and in total, we conducted 116 observations at the outlet of PYNF and TGS. The data analysis method was the same as for the study on the local scale.

Regional Scale

Plant Diversity

Yunnan province of China is recognized as a center of the origin of global plant diversity (Wu and Jiang 1987). Botanists have extensively researched the flora of Yunnan and delimited 11 floral subregions in the province (Wu and Jiang 1987, Figure 1). Long-term studies have recorded 13,850 species of seed plants over the past 150 years (nearly 98% of the estimated species for the province), which are recorded in *The inventory of seed plants of Yunnan, Flora Yunnanica* and other literature (Wu 1984, 1990; Wu and Jiang 1987; Wu et al. 1995). Moreover, ecologists have monitored soil erosion by remote sensing and GIS in the province (Zhang and Wang 2000). Thus, the Yunnan province can meet the hypothesized study conditions for a regional scale. We determined the number of plant species in each floral subregion using statistics of the distribution of each plant species in each floral subregion. Specifically, we set up electronic tables with the names of the 11 floral subregions in the first line. Then we filled in each of the 13,850 plant species' names in the first column of the table, and identified the distribution of each plant species across each floral subregion by searching *The inventory of seed plants of Yunnan* and *Flora Yunnanica.* If a species was distributed in a floral subregion, we assigned '1' to the cell corresponding to the floral subregion in the table; if not, it was assigned '0' When all 13,850 plant species were identified, the number of seed plants in each floral subregion was given by the sum at the bottom of each column. Equation (5) was used to evaluate plant diversity in these regions.

$$D_i = \frac{N_i}{A_i} \quad (i = 1,2.3\ldots\text{i}) \tag{5}$$

where D_i is plant diversity of floral subregion *i,* the number of species per km^2; N_i is species number in floral subregion *i;* A_i is the area of floral subregion *i.*

Table 3. Land area, species number and plant diversity of the floral subregions of Yunnan province, China

Floral region	A	B	C	D	E	F	G	H	I	J	K	Total
Land area (10^3 km^2)	11.96	4.31	32.62	32.68	16.94	81.63	64.41	67.44	47.18	14.63	10.32	383.21
Number of species (*s*)	3,150	769	3,962	4,368	2,994	3,720	3,085	3,602	2,607	3,597	812	32,666[a]
Plant diversity (*D*)	0.26	0.23	0.12	0.13	0.18	0.05	0.05	0.05	0.06	0.25	0.08	0.09

[a] Total number of species includes species that are common to more than one subregion. Counting these common species only once gives a species number of 13,850. Regression models of species number and plant diversity *D* on the land area of the floral subregions: species number = 0.845Ln (land area) + 0.251, $R^2 = 0.44$, $n = 11$, $F = 7.162$, $p = 0.025$; plant diversity $D = -0.07$Ln (land area) +0.3581, $R^2 = 0.60$, $n = 11$, $F = 13.765$, $p = 0.005$.

Land area, number of plant species, and plant diversity presented gradual changes among the 11 floral subregions in Yunnan province (Table 3). The number of plant species increased logarithmically with increasing area of the floral subregions, but plant diversity *D* declined logarithmically with increasing area of the floral subregions. Subregion A showed the highest diversity and subregion F the lowest, but F had the greatest land area. Plant diversity *D* was 0.09 for the entire Yunnan province.

Soil Erosion

Using ArcGIS 9.7, we superposed the borders of each floral subregion (Figure 1) on the Yunnan province maps of soil erosion from *The report about soil erosion survey of Yunnan province based on remote sense technique* (Zhang and Wang 2000). Then we calculated the area of the different grades of soil erosion and erosion amount in each floral subregion using equation (6). According to China National standards, soil erosion was classified into six grades: no obvious erosion, < 500 t km^{-2} yr^{-1}; gentle erosion, 500–2500 t km^{-2} yr^{-1}; moderate erosion, 2500–5000 t km^{-2} yr^{-1}; strong erosion, 5000–8000 t km^{-2} yr^{-1}, intense erosion, 8000–15,000 t km^{-2} yr^{-1}, severe erosion, > 15,000 t km^{-2} yr^{-1}.

$$E_i = \sum a_{ij} m_i \quad (i = 1, 2, 3\ldots i; j = 1, 2, 3\ldots j) \tag{6}$$

where E_i is amount of soil erosion for all patches of the *i*th erosion grade in a floral subregion; a_{ij} is the area (km^2) of the *j*th patch in the *i*th erosion grade in the floral subregion; m_i is the soil erosion module for the *i*th erosion grade (for no obvious erosion and severe erosion, we took 500 and 15,000 t km^{-2} y^{-1} and for other erosion grades, we took the median value between upper and lower limits). In addition, we calculated total soil erosion amount for the six erosion grades in each floral subregion using E_i. Based on D_i, E_i and the total soil erosion amount in all 11 floral subregions, we analyzed the relationship between plant diversity and soil erosion in each erosion grade.

RESULTS

Local Scale

Surface runoff, sediment and total P loss in surface runoff presented a decreasing trend with increasing plant species richness and ecosystem maturity along the successional stages (Tables 1 and 4). Specifically, the means, standard deviations, and coefficients of variation of the three indexes all became smaller, to an identical degree, with increasing species richness and ecosystem maturity. For ACKV, the most diverse and mature stage, the means, standard deviations, and coefficients of variation for the three indexes were lowest, but in the least diverse and mature stage, APMO, they were highest (Table 4). This is because the frequency of surface runoff was very low in the highly diverse plots and less runoff flowed from them. Plant density increased linearly with increasing species richness in the different layers when looking only at the relationship between the two in all plots or subplots (Figure 2A, B, and C). Total plant cover increased logarithmically with species richness (Figure 2D). In the plot with the highest measured species richness, plant cover was 86.5%, whereas it was only 29.0% in the plot with lowest species richness.

Table 4. Statistical parameters of surface runoff and sediment in the different successional stages at the local Muding site

Succession stages	df	Surface runoff			Sediment			Loss of total P		
		M (m^3 ha^{-1})	SD	CV	M (t ha^{-1})	SD	CV	M (g ha^{-1})	SD	CV
APMO	83	22.44	67.43	300.49	0.113	0.771	682.30	7.08	35.87	506.64
AEME	83	11.7	33.83	289.15	0.079	0.45	569.62	5.57	15.29	274.51
AEI	83	10.63	30.26	284.67	0.063	0.325	515.87	4.02	8.25	205.22
APLO	83	10.31	20.14	195.34	0.058	0.287	494.83	2.76	5.44	197.10
ACKV	83	8.91	16.33	183.28	0.044	0.127	288.64	1.87	3.53	188.77

df: degrees of freedom; M: mean; SD: standard deviation; CV: coefficient of variation. These statistical parameters were calculated using all 84 samples from each stage over three years (one stage reached 84 runoff events but the others had fewer runoff events; the identical sampling number of 83 is used for comparison). Because rainfall ranged from several millimeters to about 100 mm, surface runoff, sediment and loss of total P presented large standard deviation and coefficient of variation, much higher than the mean. These three erosion indexes all showed negative logarithmic correlations to species richness and linear correlations to ecosystem maturity at 0.05 confidence level when all observations of five successional stages were used to regress on species richness and ecosystem maturity.

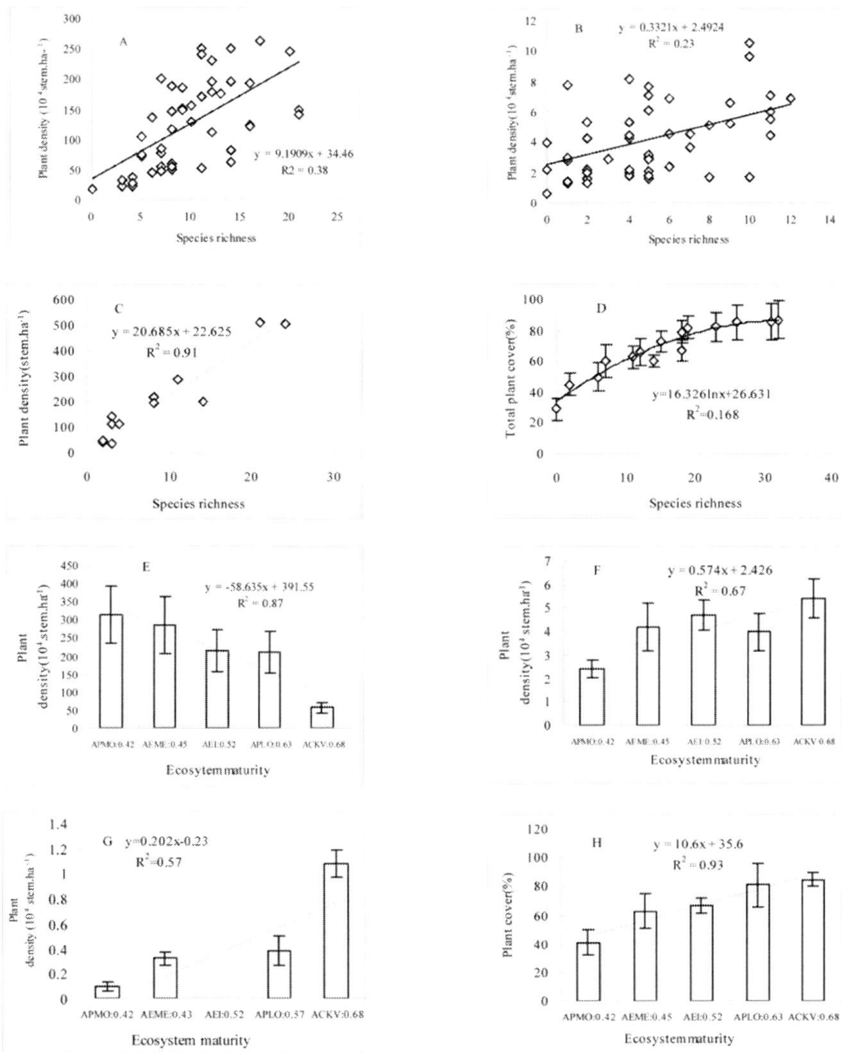

Figure 2. Relationship of plant density in herb (A), shrub (B), and tree (C) layers, and total plant cover (D) of the three layers with species richness, and relationship of plant density in herb (E), shrub (F), and tree (G) layers and total plant cover (H) with ecosystem maturity. Species richness = s_i - s_0, where s_i is the species number of plot or subplot i, and s_0 is the lowest species number among all plots or subplots. Regression models in A, B, C, D, and in E, H showed significance at 0.001 and 0.05 confidence levels, respectively, whereas regression models in F and G did not, even at the 0.05 level.

Plant density also increased with increasing ecosystem maturity in the tree and shrub layers (Figure 2F and G), but decreased in the herb layer (Figure 2E, because high ecosystem maturity coincides with high dominance in tree and shrub layers pressing density of herb plants). Total plant cover also showed an increasing trend with ecosystem maturity (Figure 2H).

Watershed Scale

The generation of surface runoff water and soil erosion in the two parts of the watershed was significantly different under similar species richness and different ecosystem maturities.

Smaller means, standard deviations, and coefficients of variation of surface runoff and sediment corresponded to the more mature PYNF (Table 5). Runoff generation in the TGS was almost fourfold that in the PYNF based on 58 observations, and soil erosion was up to 13.4 times higher in the former compared to the latter.

Table 5. Statistical parameters of surface runoff and soil erosion from all observations in the different parts of the watershed with similar species richness and different ecosystem maturity

Watershed parts	Degree of freedom	Surface runoff			Sediment		
		Mean (m^3 ha^{-1})	SD	CV	Mean (t ha^{-1})	SD	CV
PYNF	57	41.19	1392.05	3379.78	72.47	9280.10	12805.44
TGS	57	160.82	34959.67	21738.09	1245.33	1996547.32	160642.60

ANOVA: surface runoff, $F = 39.25$, $n = 116$, $p < 0.001$ between the two watershed parts; sediment, $F = 13.25$, $n = 116$, $p < 0.01$.

Regional Scale

Soil erosion for the different erosion grades declined with increasing diversity *D*. Under conditions of no obvious erosion grade or a gentle erosion grade, this decline was almost linear (Figure 3A, B). However, the relationship between soil erosion and *D* under moderate and strong erosion grades was exponential (Figure 3C, D), and a negative logarithmic relationship was observed under intense and severe erosion grades (Figure 3E, F). The total soil-erosion curve also presented the form of an exponential function (Figure 3G).

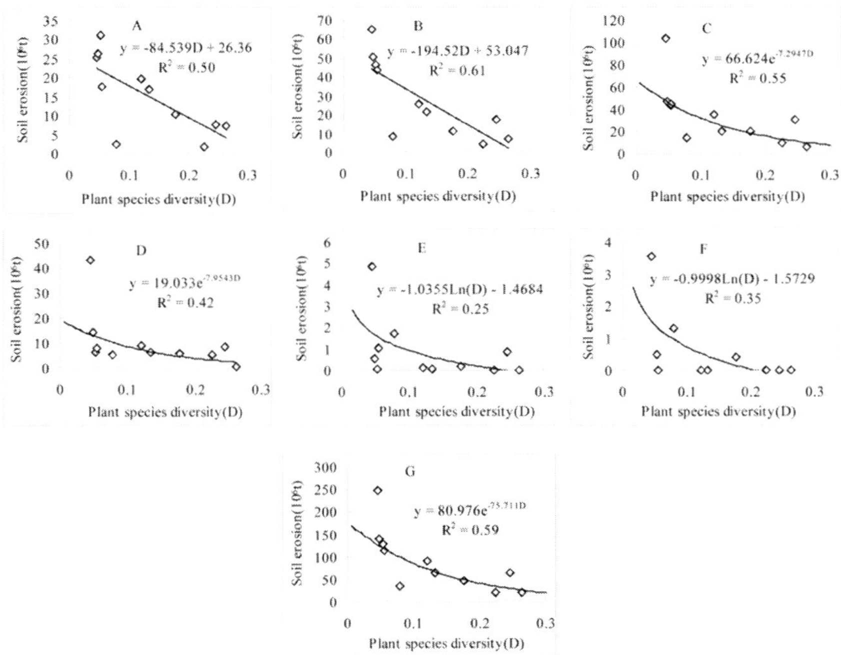

Figure 3. Relationship between plant diversity *D* and soil erosion. (A) no obvious erosion grade; (B) gentle erosion grade; (C) moderate erosion grade; (D) strong erosion grade; (E) intense erosion grade; (F) severe erosion grade; (G) total erosion. $n = 11$ (11 floral subregions). Regression models in B, C and in A, D, G showed significance at 0.001 and 0.05 confidence levels, respectively, whereas regression models in E and F did not, even at the 0.05 level.

According to soil erosion and diversity data, the floral subregions with high plant diversity corresponded with areas of mainly no or gentle erosion, whereas strong and intense erosion primarily occurred in the floral subregions with low plant diversity.

DISCUSSION

The plant diversity-ecosystem function relationship is impacted by many factors, and ecologists generally reach their different conclusions when these factors have simultaneously worked (Loreau et al. 2001). We considered the problem and designed a direct test for relationships between plant diversity and three erosion indexes under the same erosion factors, as well as indirect tests on a local scale. The significantly negative relationships between plant diversity and, respectively, surface runoff, sediment, and loss of total P, and the gradual decreases in their means, stand deviations, and coefficients of variation with increasing plant diversity indicate that plant diversity not only leads to enhanced soil conversation in these succeeding communities, but also to the stability of the soil conservation function. The positive relationship between plant diversity and plant cover and density of key layers in the study also indirectly supports the result (Lin et al. 1997; Calder 2001). Naeem et al. (1994) and Tilman et al. (1997) reported that total plant cover increases along low, intermediate and high-diversity plots, and with the number of added species in manipulated communities. Plant-assemblage experiments using control plots with increasing species richness, from 1 to 32 species, showed an increase in plant cover from 64% to 100% (Spehn et al.2000). A close positive relationship between plant diversity and density has also been found in grassland (Stevens and Carson 1999), forest (Jiang et al. 2002) and experimental (He et al. 2005) communities. These studies also support our results.

Based on our tests, we suggest that plant diversity regulates soil erosion by plant cover and density following different mechanisms. Ecosystems with diverse species composition contain more species with niche differentiation (Loreau et al. 2001). They grow to different heights and can lead to a better vertical and space-filling canopy structure (Naeem et al. 1994; Spehn et al. 2000). As a result, forest gaps decrease and plant cover increases with increasing plant diversity, which is naturally linked to enhanced soil erosion control. We define this as the covering effect of plant diversity. There is also stronger inter-species competition in highly diverse vs. low-diversity ecosystems (Tilman 1982). We conjecture that plant species tends to diffuse from strongly competitive habitats to weakly competitive ones to more easily obtain resources for growth, resulting in population immigration. Thus, species diffusion may also lead to an increase in plant cover, constituting the diffusion effect of plant diversity. In addition, stronger inter-species competition generally occurs among different populations with increasing diversity in a plant community (Tilman 1982), and these populations tend to biologically maintain a high population level so as not to fall below a minimum population size that could be excluded from the community (Mode 1988). Thus, each species may tend to develop adaptive ecological strategies to maintain a greater population, and plant density increases. This process represents the density effect induced by plant diversity.

We expected that surface runoff, soil erosion and their standard deviations and coefficients of variation would not be obviously different between PYNF and TGS at the watershed scale, which would show that plant diversity has a larger effect on soil erosion control than ecosystem maturity. However, this was not the case, indicating that ecosystem maturity plays a greater role in soil erosion control than plant diversity. However, watershed ecosystems consist of local ecosystems, and erosion control at the watershed scale must be influenced by the relationship between plant diversity and erosion control on the local scales. Highly diverse local sites should have weaker erosion than low-diversity sites in PYNF and TGS. The strong process of "ecosystem maturity-erosion control" in PYNF and TGS might overshadow the local effects of plant diversity. Thus, we suggest that plant diversity has a weaker effect on soil erosion control than plant density, key species, plant cover or ecosystem maturity (Guo et al. 2006)

We found a negative correlation between plant diversity and soil erosion on the regional scale. In this study, soil erosion grade was based on plant cover and slope degree of each erosion patch extracted by GIS; thus soil erosion data were dependent on plant cover and slope degree, and soil erosion volume was positively related to slope degree and negatively related to plant cover (Zhang and Wang 2000). These studied processes indicate that decreases in soil erosion with increasing plant diversity in floral subregions (Figure 3) must be due to the increase in plant cover with increasing plant diversity. Therefore, the covering effect and diffusion effect on the local scale can partially explain the response of soil erosion control to plant diversity on the regional scale. A reasonable explanation for the relationship on this scale is that floral subregions with high diversity are well conserved regional ecosystems with high plant cover and biomass; as a result, soil erosion control is excellent. The plant species data used in this study for the 11 floral subregions have been accumulated for over 100 years, and the soil-erosion data are generally acknowledged. These results largely describe the responses of soil erosion to changes in plant diversity on a regional scale.

In summary, our study, performed on three scales, describes the effects of plant diversity on soil-erosion control. Previous studies have emphasized the effects of plant cover, litter and biomass. This work provides a new understanding of plant diversity as a vegetation factor indirectly regulating soil erosion. Because soil erosion on different scales impairs primary productivity of ecosystems, soil-erosion control in diverse ecosystems provides good nutritional conditions for plant growth. Therefore, plant diversity can maintain ecosystem productivity. This study therefore supports the diversity-productivity relationship by "plant diversity-soil conservation-primary productivity" in a roundabout way.

Acknowledgments

This work was supported by grants from the National Natural Science Foundation of China (40861015), and the Foundation of State Key Laboratory of Soil Erosion and Dry-land Farming on the Loess Plateau (No. 10501-208).

References

Adler PB, Seabloom EW, Borer ET, Hillebrand H, Hautier Y, Hector A, Harpole WS, O'Halloran LR, Grace JB, Anderson TM, Bakker JD, Biederman LA, Brown CS, Buckley YM, Calabrese LB, et al. 2011. Productivity is a poor predictor of plant species richness. *Science,* 333:1750–1753.

Calder IR. 2001. Canopy processes: implications for transpiration, interception, and splash induced erosion, ultimately for forest management and water resource. *Plant Ecology,* 153:203–214.

Chapin FS, Zavaleta ES, Eviner VT, Naylor RL, Vitousek PM, Reynolds HL, Hooper DU, Lavorel S, Sala OE, Hobbie SE, Mack MC, Díaz S. 2000. Consequences of changing biodiversity. *Nature,* 405:234–242.

Costanza R, d`Arge R., de Groot R, Farber S, Grasso M, Hannon B, Limbur K, Naeem S, O`Neill RV, Paruelo J, Raskin RG, Sutton P, van den Belt M. 1997. The value of the world's ecosystem services and natural capital. *Nature,* 387:253–260.

Feng ZW. 1999. Biomass and Productivity of Forest Ecosystems in China. Beijing, China: Science Press.

Guo QF, Shaffer T, Buhl T. 2006. Community maturity, species saturation and the variant diversity–productivity relationships in grasslands. *Ecology Letters,* 9:1284–1292.

He JS, Wolfe-Bellin KS, Schmid B, Bazzaz FA. 2005. Density may alter diversity-productivity relationships in experimental plant communities. *Basic and Applied Ecology,* 6:505–517.

Hector A, Schmid B, Beierkuhnlein C, et al. 1999. Plant diversity and productivity experiments in European grasslands. Science 286:1123–1127.

Hector A, Schmid B, Beierkuhnlein C, Caldeira MC, Diemer M, Dimitrakopoulos PG, Finn JA, Freitas H, Giller PS, Good J, Harris R, Högberg P, Huss-Danell K, Joshi J, Jumpponen A 2000. No consistent effect of plant diversity on productivity. Science 289:1255–1256.

Jiang HQ. 1984. Several issues in researching about *Pinus yunnanensis Journal of Yunnan University,* 6:1–5.

Jiang YX, Wang BS, Zang RG, Jin JH, Liao WB. 2002. Biodiversity and Formation Mechanism of Tropical Forest in Hainan Island in China. Beijing, China: Science Press.

Lal R. 1988. Soil Erosion Research Methods. New York: Soil and Water Conservation Society.

Li ZJ. 2008. Effects of Vegetation Restoration on Physical and Chemical Properties of Soil Profiles in Distribution Area of Purple Soil in Central Yunnan Province [Master's thesis]. Kunming, China: Yunnan University.

Lin TC, Hamberge SP, King HB, Hsia, YJ. 1997. Spatial variability of through-fall in the subtropical rainforest in Taiwan. *Journal of Environmental Quality,* 26:172–180.

Loreau M, Naeem S, Inchausti P, Bengtsson J, Grime JP, Hector A, Hooper DU, Huston MA, Raffaelli D, Schmid B, Tilman D, Wardle A. 2001. Biodiversity and ecosystem function: current knowledge and future challenges. *Science,* 294:804–808.

Mode CJ. 1988. On statistically assessing critical population size of an endangered species in a random environment. *Mathematical Medicine Biology,* 5:147–166.

Morgan RP. 1996. Soil Erosion and Conservation. London: Group UK.

Naeem S, Thompson LJ, Lawler SP, Lawton JH, Woodfin RM. 1994. Declining biodiversity can alter the performance of ecosystems. *Nature,* 368:734–737.

Ricklefs RE. 2008. Disintegration of the ecological community. *The American Naturalist,* 172:741–749.

Schulze ED, Mooney HA. 1993. Biodiversity and Ecosystem Function. New York: Springer-Verlag.

Spehn EM, Joshi J, Schmid B, Diemer M, Körner C. 2000. Aboveground resource use increases with plant species richness in experimental grassland ecosystem. *Functional Ecology,* 14:326–337.

Stevens MH, Carson WP. 1999. Plant density determines species richness along experimental fertility gradient. *Ecology* 80:455–465.

Szwagrzyk J, Gazda A. 2007. Above-ground standing biomass and tree species diversity in natural stands of Central Europe. *Journal of Vegetation Science,* 18:555–562.

Tang CQ. 2010. Subtropical montane evergreen broad-leaved forests of Yunnan, China: diversity, succession dynamics, human influence. *Frontiers of Earth Science in China,* 4(1):22–32.

Tang CQ, Hou X, Gao K, Xia T, Duan C, Fu D. 2007. Man-made versus natural forests in mid-Yunnan, southwestern China: plant diversity and initial data on water and soil conservation. *Mountain Research and Development,* 27(3):242–249.

Tilman D. 1982. Resource Competition and Community Structure. Princeton, NJ: Princeton University Press.

Tilman D. 2000. Causes, consequences and ethics of biodiversity. *Nature* 405:208–211.

Tilman D, Knops J, Wedin D. 1997. The influence of functional diversity and composition on ecosystem processes. *Science,* 277:1300–1302.

Tilman D, Wedin D, Knops J. 1996. Productivity and sustainability influenced by bio-diversity in grassland ecosystems. *Nature,* 379:718–720.

William JR, Renard KG, Dyke PT. 1983. EPIC—a new method for assessing erosion, effect on soil productivity. *Journal of Soil and Water Conservation,* 36:381–383.

Wang, L.X. 1997. Soil and Water Conservation. Beijing, China: China Forestry Publishing House.

Wu CF. 2003. Land Ecology. Beijing, China: China Land Press.

Wu ZY. 1984. The Inventory of Seed Plants of Yunnan. Kunming, China: Yunnan People Press.

Wu ZY. 1990. Flora Yunnanica. Beijing, China: Science Press.

Wu ZY, Jiang HQ. 1987. Yunnan Vegetation. Beijing, China: Science Press.

Wu ZY, Wang XF, Zhu YC, Zhang XS. 1995. China Vegetation. Beijing, China: Science Press.

Xu XY. 2002. The Change of Soil Ecological Fertility of Degraded Mountain Ecosystem in Different Periods of Restoration in Middle Yunnan Plateau [Master's thesis]. Beijing, China: Yunnan University.

Zhang F, Wang L. 2000. The Report about Soil Erosion Survey of Yunnan Province Based on Remote Sense Technique. Kunming, China: Yunnan Water Resources Bureau.

In: Forest Ecosystems
Editor: Noel C. Roberts

ISBN: 978-1-63117-815-3

Chapter 6

HYDROLOGICAL EFFECT OF FOREST LOGGING IN BOREAL ZONE OF SIBERIA

T. A. Burenina**, *A. A. Onuchin and E. V. Fedotova
V. N. Sukachev Institute of Forest Siberian Branch
the Russian Academy of Sciences,
Krasnoyarsk, Russia

ABSTRACT

Siberian rivers account for a great part of water discharge into the Arctic Ocean. Along with the regional climate and the presence permafrost, hydrological processes occurring in catchments of the boreal taiga rivers are controlled by forest harvesting levels. Forests of the Angara region and West Sayan have been harvested extensively over the past fifty years. As a result, forest area has been reduced and forest age structure and composition have changed. These changes are reflected in the natural water budget structure and hydrological regimes of areas. The study revealed that hydrological effects of forest logging conducted in North Angara region characterized by a highly continental climate and, hence, severe forest growing conditions differ from those in West Sayan.

---

* E-Mail: burenina@ksc.krasn.ru.

INTRODUCTION

Temperate forests covering 34% of the Earth's land have experienced human disturbances since ancient times and this is where the major industrial and agricultural enterprises are located nowadays. The heaviest human stress has been on European and North American ecosystems. In Europe, extensive agricultural activities, gardening, forest planting, and development of towns and villages have resulted in practically complete elimination of virgin forests from this region. As much as ¾ of all forests have been completely destroyed in China. The USA has lost 1/3 of their forests, with virgin forests accounting for 85% the total loss. In the only two countries, Russia and Canada, which still contain large, relatively undisturbed forests totaling 1.4 billion hectares, forest areas decrease annually by 2-3 million hectares [10].

Forest ecosystems are a major control of moisture circulation in watersheds. Boreal forests have been long recognized to influence global moisture circulation considerably. Therefore, it is logical to expect that further large-scale forest cover disturbances, such as catastrophic fires, large-scale insect outbreaks and forest logging, will affect the processes of water resource development and hydrological regime in any river basin.

Of all human activities, forest logging and post-logging regeneration have the greatest influence on water budget and total runoff from watersheds [3, 16, 21, 21, 27]. Reported first-year increases in water yield following forest clearance in the humid tropics, for instance, range from 110 to 825 mm, depending on local rainfall [14]. A dated, but still valid review of almost 100 paired catchments experiments throughout the world [3] indicated that all of those involving removal of forest cover resulted in higher stream flow totals.

The association of flooding with forest logging remains the most disputable issue of forest hydrology. Hewlett [17] reviewed the evidence from watershed research worldwide and reported that no cause and effect was demonstrated between forest cutting in the headwaters and floods in the lower basin. The disastrous floods of November 1988 in southern Thailand were mistakenly blamed on logging, and the government issued a nationwide ban on logging [37]. The same year, a flood disaster in Bangladesh was erroneously attributed to deforestation in the Indian and Nepalese mountains, where the mighty Ganges and Brahmaputra rivers have their headwaters. This was international scapegoat to avoid tough decisions about floodplain occupancy and river management in the lower basin [15].

A number of studies [42, 44, 45] showed that fire- or logging-caused conversion of forest lands to grasslands only leads to local flood increases and is not among the key controls of large-scale high water. A publication by FAO and the Center for International Forestry Research [10] has this to say: although forests can play a certain role in delaying and reducing peak floodwater flows at local levels, scientific evidence clearly indicates that forests cannot stop catastrophic large-scale floods, commonly caused by severe meteorological events.

Numerous investigations on forest hydrology conducted in different parts of Siberia show that clear cuts cause deep and persistent changes of water regulating and soil conserving functions of forest ecosystems [4, 5, 6, 24, 25, 29, 31, 32].

Many rivers of southern taiga and mountain regions of Siberia drain areas of land that experienced a dramatic land cover change, with a decrease in overall forest area and a relative increase in deciduous trees. Land cover change in forest catchments (cutting down, wild fires) impact on water balance and water-protective functions of forest. Forest destruction results in increasing soil erosion, decreasing water quality in still-water bodies and streams, undesirable runoff changes, and local and large-scale hydrological regime disturbances.

The regional water resource response to changing watershed forest area and condition, as well as land uses, is of special interest. In this respect, investigating forest-harvesting impacts and post-logging forest recovery on water balance of territory and forest water protection and erosion prevention functions is critically important from a practical viewpoint.

Rivers found in the Angara-Yenisei catchments were studied for runoff dynamics. This river runoff is formed in the southern taiga forest zone and mountains of Southern Siberia. Formation of Siberian rivers runoff and its season dynamics depends on forest vegetation to a considerable extent. Forest vegetation, along with climate, soil, topography, and lithosphere characteristics, has a considerable influence on moisture budget structure and surface and ground water quality. Numerous studies conducted in different climatic and vegetation zones of Siberia revealed a significant impact of forest ecosystems on river runoff amount and quality. In this respect, forests are important in terms of snow distribution and accumulation, as hydrological regimes of most Siberian rivers are determined mainly by watershed snow moisture budget.

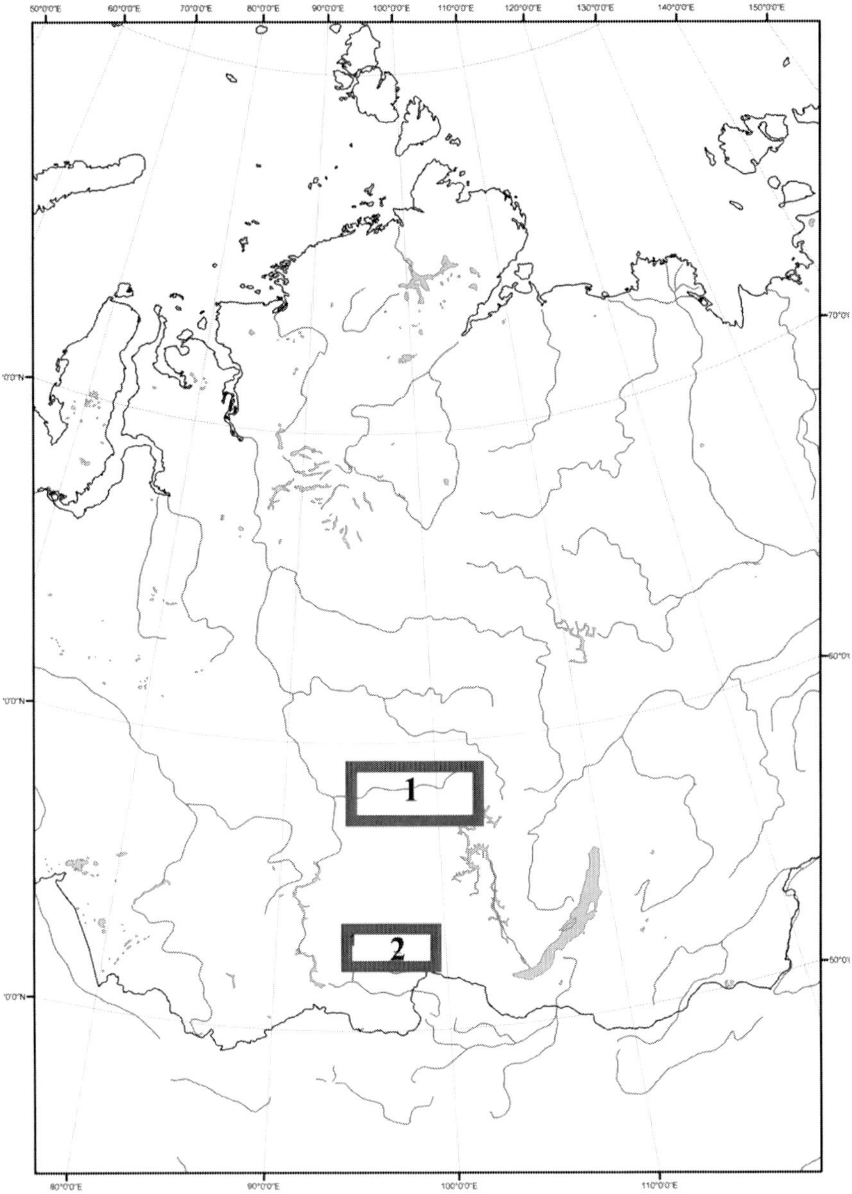

Figure 1. Study area. 1- North Angara region, 2 – West Sayan Mountains.

REGION OF INTEREST AND STUDY AREAS

Our study was conducted in forest ecosystems and watershed found in Western Sayan Mountains and North Angara region; these areas are situated in Central Siberia (Figure 1).

Sayan Mountains

West Sayan Mountain Ridge, a wide strip of mountains with elevations above sea level (a.s.l.) of up to 2500 m, stretches from Abakan river head eastward to Kazyr, Uda, and Kizhi-Khem river heads in East Sayan Mountain Ridge. In the north, West Sayan borders on vast hilly Minusinsk Hollow covered by forest-steppe vegetation and central Tuvan Hollow occupied by dry steppe is adjacent to this mountain ridge in the south. The regional commercial woody species in West Sayan, of which the most valuable are Siberian pine (*Pinus sibirica*) and Siberian fir (*Abies sibirica*) constitute a wide altitudinal belt of dark-needled conifers. This region is in moderately continental climate characterized by mild winter and warm summer seasons. Average annual air temperature is minus 1°C and precipitation ranges 600 mm to 1200 mm depending on elevation a.s.l. In foothills, snow pack is 60-80 cm deep, whereas snow depth increases to 100-180 cm, depending on slope aspect, at high elevations.

All the altitudinal vegetation belts (AVB), from forest-steppe to mountain tundra, common in the mountains of southern Siberia can be found on the northern macro slope of West Sayan. A chern taiga belt located between 300 m and 700 m a.s.l. is higher replaced by the mountain taiga forest belt, which becomes fairly open at 1200-1300 m a.s.l. and is replaced by sub alpine Siberian pine/Siberian fir open woodlands and alpine meadows. Between 1400 m and 1500 m a.s.l., these open woodlands are replaced by tundra vegetation.

North Angara Region

The study area was a part of Angara river catchments located in Central Siberian Plateau. This is a rough terrain (av. 310 m a.s.l.) having no pronounced upland sites and characterized by extremely continental climate with an average annual air temperature of minus 4°C and precipitation ranging 312 mm to 355 mm. Normally, snow cover becomes stable in late October, with the earliest snow cover stabilization observed in early October. On the right bank of Angara river snow layer depth averages 75-76 cm, whereas the layer is only 45-46 cm deep on the left bank, because this area receives little solid precipitation. The study area contains various types of soil, which are seasonally deep-frozen due to long, cold winter. The area falls within the vegetation zone of southern taiga mixed larch/Scots pine forests, with dark conifers, such as spruce and fir, limited to river valleys and small flat-bottom ravines. Deciduous forests are common on old burns and logging sites.

Monitoring water regime changes in Central Siberia is a high research priority due to ever-growing forest logging and water resource use. West Sayan and North Angara region vegetation cover has been changed considerably by logging over several past decades, and these changes influenced water balance structure of disturbed sites. Accessible low and middle mountain forest stands were largely clear cut in 20 century and they are now represented by various forest regeneration stages. Intact stands of the regional major woody species are found only in inaccessible areas. Therefore these regions are nowadays a mosaic of forest regeneration sites including both post-human and post-fire regeneration patterns.

METHODS

Hydrological Parameters Estimation

This study covered forest ecosystems and watersheds in different parts of central Siberia to include Western Sayan Mountains and North Angara region. We used our multiyear forestry and hydrological data obtained at the Institute of Forest field stations, as well as the results of snow surveys in different regions of Siberia. Information contained in climate reference books, hydrology year-books, and literature was also analyzed. Certain watersheds have been observed for over twenty years. Such long periods of observation allowed us to cover a wide range of hydroclimatic situations.

The *in situ* data on moisture evaporation from snow and soil surface combined with experimental and calculated data on moisture used for transpiration and on evaporation of intercepted precipitation obtained for different parts of Siberia [5, 12, 19, 20, 24, 25, 28] were used to calculate the total moisture evaporation.

Calculating evaporation of intercepted precipitation used our data on foliage biomass in both conifer and deciduous species found in Siberia [30, 34, 46] and information on understory vegetation biomass reported by other authors [1, 2]. Air temperature and precipitation information, river runoff characteristics were studied using reference books [18, 37, 41] and data of Hydro meteorological centre of Krasnoyarsk kray [47]. Changes in forest logging intensity and wildfire activity were estimated based on satellite imagery and forest inventory data analyses. Observation periods lasted for over fifty years (1946-2002) to cover a range of hydro climatic conditions for some of the study sites.

For the catchments of interest, which are characterized by complex spatial vegetation patterns, total moisture evaporation and runoff were calculated as weighted mean values:

$$Z = \Sigma Z_i f_i / 100; \; Y = \Sigma Y_i f_i / 100 \quad (1)$$

where *Z* and *Y* are landscape-specific total moisture evaporation and runoff; *f* is corresponding landscape percent areas.

Landscape Analysis Using Digital Elevation Model and Remote Sensing Data

To extrapolate the hydrological observation results obtained on experimental plots to AVBs areas and the river catchments, we use Landsat images obtained for different years since 1974 and a digital elevation model (DEM).

The landscape structure was analyzed using DEM in SRTM Version 4 format. The West Sayan study area was divided into three elevation levels: up to 700 m, 700-1400 m, and over 1400 m a.s.l. and into six slope classes: (1) 0-1°; (2) 2-5°; (3) 6-7°; (4) 8-15°; (5) 16-25°; and (6) over 25°. Slope aspects were described using a scale of eight units. The landscape units identified encompassed various slopes containing all AVBs, from chern taiga to mountain tundra. Also, DEM helped identify Kebezh and Taigish river catchments.

Table 1. Logging sites classification accuracy

Year	Total classification accuracy, %	Kappa coefficient
1974	93	0,68
1976	94	0,76
1989	91	0,74
1991	88	0,66

Assessment of the dynamics of the logging sites in the study area involved the use of the 1974, 1976, 1989, 1991, 1999, 2001, and 2010 Landsat images (http://glovis.usgs.gov). All the images were geo-referenced to the 2010 image with an error near of one pixel. This operation was done to improve spatial image coincidence. As the study area is a mountain area, the images were topographically normalized using C-factor method [39]. The normalization quality was assessed by selecting, in each image, five fragments to cover sunlit and shaded elements and calculating variation coefficients for the pixel values found within these fragments in both original and normalized images. The coefficients appeared to be lower for all normalized images compared to the originals to indicate a decrease in contrast between sunlit and shaded fragments.

To classify the 1974, 1976, 1989, and 1991 images, learning samples were created for logging sites of various ages. The classification was done by a method of parallelepiped, with the maximum likelihood approach being used where overlaps occurred. The classification accuracy was evaluated using an error matrix and kappa statistics [40]. The classification appeared to be av. 91% accurate for all images (table 1).

The 1999 image was used to assess the areas of the dominant vegetation cover types with account of their altitudinal location. Within each AVB, dark-needled conifer stands, deciduous forest, grasslands, bogs, and bare soil areas were identified.

Regarding the 2010 image, the task was to divide logging sites of all the ages considered into three vegetation cover types: (1) conifer forest; (2) deciduous forest; and (3) grasslands. The former logging sites already covered by forest corresponded to various forest regeneration stages, whereas the logging sites that have become grasslands were considered as sites where forest has not started to regenerate yet. Also, training samples were created for these types without involving logging site fragments. This image was classified using the minimum distance method.

RESULTS AND DISCUSSION

Numerous investigations on forest hydrology [13, 24, 25, 32] conducted in different parts of Siberia show that clear cuts cause deep and persistent changes of water regulating and soil conserving functions of forest ecosystems. From ten to forty percent of precipitation received by a forest area is intercepted by tree crowns and evaporates before reaching the ground surface. Vegetation cover changes are known to result in redistribution of moisture proportions at watersheds. Clear felling, for example, results in increasing the amount of precipitation reaching the soil surface leading to an increase in surface flow. Human-caused alterations of watershed vegetation and soils induce changes in runoff genesis, even hydro climatic conditions remaining the same. Principal felling results in reducing areas of mature forests dominated by major conifer species and increasing young conifer and secondary deciduous forest proportions. These forest cover changes influence hydrological regime, particularly the total moisture evaporation and runoff. The water balance of a forest area is also controlled by post-logging forest succession trajectories. Our long-term investigations conducted in Central Siberia allowed identifying forest-induced changes of local and regional hydrological balance.

Dynamics of Forest Cover under Human Activity

West Sayan. In West Sayan accessible mountain forest stands were largely logged in 1930-1960 years and they are now represented by various forest regeneration stages. Intact stands of Siberian Pine (*Pinus sibirica*) are found only in inaccessible areas.

In West Sayan, the ability of dark conifer forest ecosystems to regenerate after logging is controlled by their types and elevations of location [9, 35]. Fir and mixed Siberian pine-fir forests found within the taiga mountain zone regenerate without tree species conversion. Deciduous tree species occupy only fairly small mineral-soil patches of logged sites. Since fir begins to dominate as early as during the first post-logging tree generation establishment, its stands recover fully after the disturbance in this zone.

In the chern taiga zone found immediately below the taiga mountain zone, secondary stands, which replace those of the major tree species as a result of logging, are of heterogeneous species composition [23]. Logged Siberian pine stands are replaced in this zone either by pure birch, or by pure aspen, or mixed conifer/deciduous stands. This indicates that post-logging forest regeneration trends vary here, although they take place in homogeneous soil and topographic conditions.

We assessed the areas logged in each decade between 1926 and 1995 (Figure 2) using the past forest inventory information available for West Sayan. The diagram obtained shows that the biggest area logged on the northern macro slope of West Sayan occurred in the mid 1930s. As large-scale logging was started here in early 1930s, we suggested that over 1000 ha of forest were annually subjected to harvesting by mid 1930s. Between 1945 and 1965, the annual area logged in dark conifer taiga dropped down to 400-500 ha. After 1965, by which time most easily accessible forest sites of the macro slope had been logged, the annual area logged dropped even lower to become 100 ha. This situation continued to 1985, when the total area logged increased due to the introduction of new harvesting technologies that enabled large-scale wood extraction from mixed Siberian pine/fir stands found on steep slopes and at higher elevations of the taiga mountain zone.

Remote sensing data were based upon to assess the level of West Sayan forest cover disturbance by large-scale forest harvesting in the last quarter of the 20th century. Vegetation cover of Kebezh and Taigish river basins was classified using 1999 Landsat image. Evaluating the influence of forest harvesting on the regional water balance and runoff required determining areas and locations of the major conifer species, secondary deciduous stands, and treeless sites found within the altitude vegetation belt (AVBs), each of which contributes to river runoff (table 2).

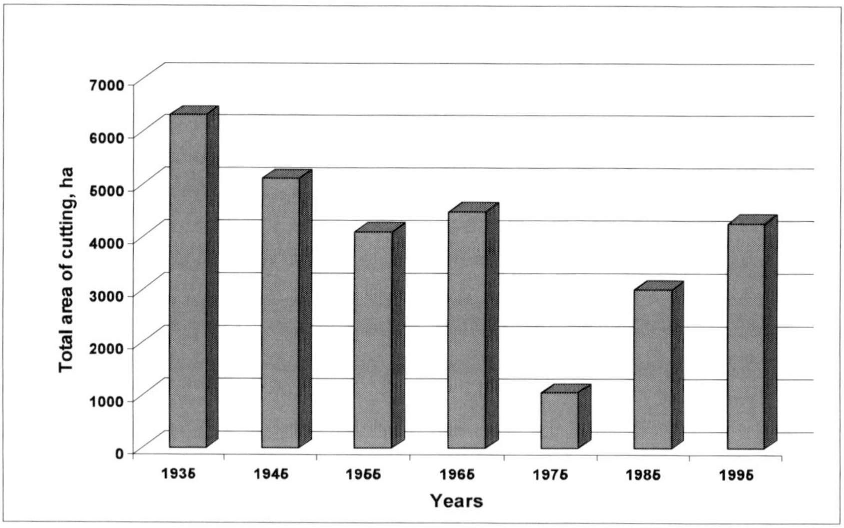

Figure 2. The 1935-1995 total area logged in the taiga forests of West Sayan Mountains on a decade basis.

The landscape structure of the studied watersheds is determined by their geographical location and features of the orography. Landsat images analysis showed in the mountainous part of the Kebezh River basin the chern taiga zone occupies 53%, mountain taiga almost 43%, and tundra and sub alpine zones only 4% of the area. In the mountainous part of the Taigish River basin, the chern taiga and mountain taiga zones cover 44 and 38%, respectively, while more than 18% of the watershed area is within the alpine-tundra and sub alpine zones.

Analysis of the landscape differentiation of the studied area shows that the spatial structure of vegetation is determined by the specifics of the reforestation process on logged sites in chern taiga and mountain-taiga zones. The largest mosaic structure of vegetation cover is typical of the chern taiga belt, since the process of reforestation in logged sites in the chern taiga has variable types. Forming stands are characterized by great diversity in respect to species composition, age, and other forest taxation characteristics, which were noted in the course of field studies [9, 23].

The result of the analysis of remote sensing data revealed that over 40% of the total area of the Kebezh River basin and about 35% of the total area of the Taigish River basin is covered by secondary forests, which reflect the degree of disturbance of the virgin forests in these basins. Forests of the chern taiga zone are characterized by the greatest disturbance: more than 65% of its area is covered by forests of post logging origin. In the mountain taiga zone, the part of secondary forests accounts for 15–17%.

Landsat image of the 2010 year was analyzed to show actual land caver. Two classes were of special interest: grassland and deciduous stands.

At up to 500 m a.s.l., grasslands corresponded to agricultural lands, between 500 m and 1200 m a.s.l. - non-forested logging sites, and the elevation over 1200 m a.s.l. - alpine meadows. Spatial analysis of the classification results and DEM revealed that non-forested logging sites were largely located on southern slopes of 7-23° between 750 m and 1125 m a.s.l. These sites were harvested in 1991 (they were dark-needled conifer forest sites in images obtained before 1992).

The 1995 forest inventory data show that fourteen logging sites occurred in 1990-1991 and include, apart from actually logged sites, forest glades and burns. The logging site class also appeared to contain two young birch sites, because young birch trees differ little from grass communities in spectral characteristics. Sites of young fir (up to 40 years of age) found on slopes of 3-15° between 800 m and 1200 m a.s.l. also fell within the logging site class, because young fir stand spectral portraits are similar to those of grass communities.

Territory really corresponding to non-forest class in the grassland class was evaluated to total 74%. The estimated total area of logging sites having no forest cover is thus 1700 ha across the two river catchments under study. These sites are located mostly at Kebezh river head and account less than two per cent of the total catchments area.

Table 2. Classification of the earth's surface of catchments of Kebezh and Taigish rivers

Altitudinal vegetation belts	Landscapes	Kebezh		Taigish	
		Area, ha	% of watershed's area	Area, ha	% of watershed's area
Mountain tundra	Rock outcrops	1550	1,6	1560	2,2
	Dwarf birch (*Betula nana*) sites (yerniks)	480	0,5	710	1,0
	Grass-bog sites	290	0,3	410	0,6
Subalpine	Rock outcrops	500	0,5	7680	11,0
	Tall and small shrubs	440	0,4	810	1,2
	Dark-needled conifer open woodlands	880	0,9	1140	1,6
	Alpine meadows	200	0,2	280	0,4
	Grass-bog sites	-	-	210	0,3
Mountain taiga	Rock outcrops	1530	1,6	1340	1,9
	Dark-needled conifer open woodlands	1080	1,1	950	1,4
	Undisturbed dark-needled conifer forest	31810	32,5	19450	27,8
	Secondary mixed conifer/deciduous stands	6350	6,5	4060	5,8
	Logged land and fire scars	710	0,7	570	0,8
	Grass-bog sites	320	0,3	220	0,3
Chern taiga	Rock outcrops	410	0,4	310	0,4
	Open woodlands and shrubs	260	0,3	270	0,4
	Undisturbed dark-needled conifer stands	15390	15,7	8090	11,6
	Secondary mixed conifer/deciduous forest	32360	33,0	18990	27,2
	Logged land and fire scars	1840	1,9	1600	2,3
	Grass-bog sites	1590	1,6	1220	1,7
Total		97990	100,0	69870	100,0

Table 3. Logging sites area

Year	Accumulated area, ha
1974	4550
1976	5555
1989	13550
1991	14365
1999	15735
2002	15760

The next step was to analyze image fragments corresponding to deciduous forest class. These were logging sites containing secondary forest stands. The logging sites having young secondary stands were largely found on north-facing slopes of 9-19° between 450 m and 850 m a.s.l. Landsat imagery since 1974 was used for a more detailed analysis of post-logging forest cover changes. As is clear from Table 3, the total logging site area was extended by about 1000 ha in 1974-1976, 8000 ha from 1976 throughout 1989, and 800 ha between 1989 and 1991. Very little forest harvesting has been done since 2000. The area of forest harvested in the mountain parts of Kebezh and Taigish catchments before 2010 was estimated to total over 15700 ha.

The logging site contours were compared with the regional landscape contours to indicate that between 1974 and 1976 the lower (chern taiga) AVB was subject to intensive forest harvesting, whereas by 1989 harvesting activities were moved upward, to mountain taiga AVB, which shift is well-pronounced in the 1991 and 1999 images. Many logging sites were limited to opposite banks of rivers on fairly steep (8-25°) "west"- and "east-facing" slopes of the north-facing West Sayan macro slope. The contours of the logging sites found within Tanzybei forest management area were then superposed onto the 1995 forest inventory map and the information related to the elementary forest inventory units (EFIU) encompassed by these contours was retrieved. Of the total area of the logging sites identified in the 1974 image, 900 ha (half of all EFIUs) was accounted by 20-30-year-old mixed birch/aspen/fir/Siberian pine stands. Dark-needled conifer stands aging 15-20 accounted for 70% of all EFIUs (196 ha) covered by logging sites according to the 1976 image. In the 1989 image, 80% of all EFIUs (2200 ha) within logged sites were covered by mixed 10-25-year-old birch/aspen/fir/Siberian pine stands. The respective values were 90% and 620 ha (20-25-yr-old fir stands) for logging sites identified in the 1992 image. This analysis revealed a pronounced difference in post-logging forest succession between chern taiga and mountain taiga AVBs: the chern taiga logging sites of 1974-1976 appeared to recover through mixed stands, whereas the 1991 logging sites in mountain taiga contained fir- or Siberian pine-dominated dark-needled stands.

These data on logging-induced structural forest cover changes combined with the results of the investigation of the regional hydrological regime allowed us to analyze spatial and temporal water balance variability depending on forest succession trajectories.

North Angara region. Forest harvesting is a historical human activity in Angara region, the environmental effects of which have ranged from positive to catastrophic. This variability of logging effects is largely due to poorly thought-out regional land use strategies. Easily accessible commercial forest sites already disturbed by previous economic activities have been clear cut using ecologically unacceptable harvesting methodologies and heavy machinery.

The forest cover of North Angara region has changed considerably over the past several decades as a result of intensive wood extraction, wildfire, and biotic influences. Large-scale wood extraction started in this region in early 1960s and totaled as many as over 10 million cubic meters by 1990. According to Sokolov et al. [41], total area logged here by 1988 was 2.4 times greater as compared with 1966. This was the time period when about 500 thousand ha of forests, including 350 thousand ha of Scots pine, were covered by clear cutting.

Most of the accessible forest stands were subject to principal cutting in 1990s and now these logging sites are a mosaic of forest regeneration stages including young stands of fire origin. Undisturbed stands of the major woody species can now be found only in Angara-side area or far, 100 km and more, from Angara river.

The regional forest condition is considerably controlled by human activities, which are a key factor accounting for changes of the major woody species proportions. As a result of a harvesting forest area has been reduced and forest age structure and composition have changed. Three types of regenerations take place in Northern Angara region: dark-needled stands replace secondary wood; pine and larch stands on loam soils replace mixed pine-birch wood; pine and larch stands on sand and loamy sand soils – without vegetation change.

The catchments of the second-order tributaries of Angara river were analyzed for logging- and fire-caused disturbances using the 2002 space images and 1:200 000 digital topographic map. The analysis covered 28 catchments ranging 20 to 50 km^2, 16 catchments ranging 50-100 km^2, and eight catchments each over 100 km². For the catchments of 20-50 km², sites disturbed by logging and fire accounted for 30-50% of the total catchments areas; including 5-10 % freshly logged sites and the remaining portions were under young stands. For larger catchments disturbed by logging and fire accounted for 15-20% of the total catchments areas by 2002; 10-15% was under young stands and not more than 5% under freshly logged sites.

We used data kindly provided to us by V.V. Kuzmichev to analyze the 1950-1990 structural changes of standing crop for four Forest Management Areas (FMA) located in Angara region. The inventory forestry data of 1950- and 1990-years were compared (figure 3). According to the figure 4 portion of Scots pine stands in total timber volume reduce from 50-60% to 35-40%; a portion of birch and aspen wood increases 2-3 times.

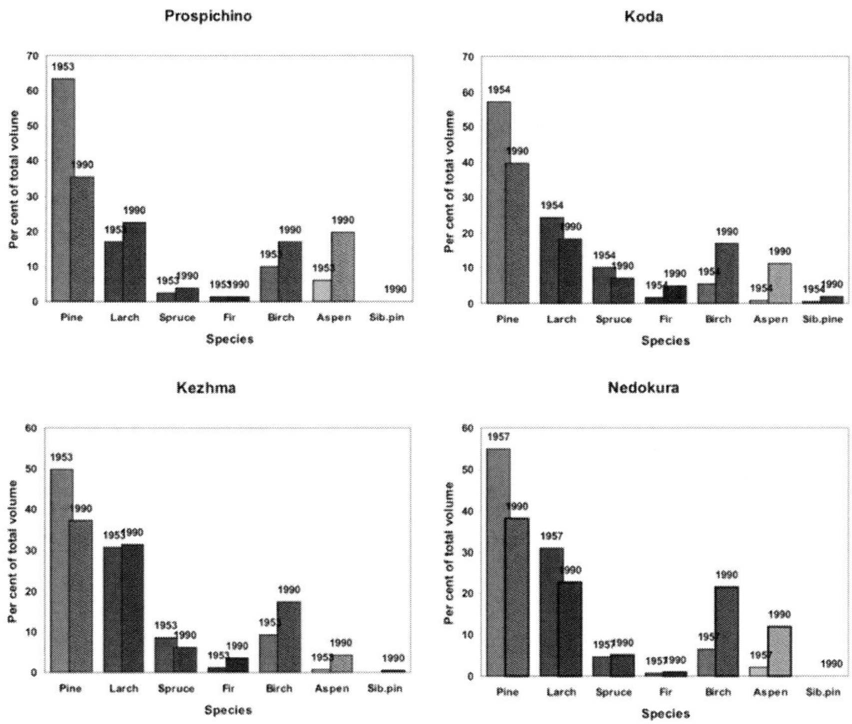

Figure 3. Timber volume structure dynamics in North Angara Region forest.

As is clear from the visual analysis of the 2011 Landsat image for this area, no forest harvesting is now being done here, because the commercial wood resources have been exhausted. Prospikhino and Nedokura sites contain no conifer forest and it contributes as little as only about 30% to the forest in Koda site. Besides logging the forest cover is highly disturbed by fairly recent wildfires.

We used the results of our study and literature information [11, 41] to assess the forest disturbance rate for the entire Angara Region and to detail post-fire and post-logging forest regeneration. The diagram in Figure 4 shows averaged disturbance data for the regional major tree species. We classified stands, which originated as a result of long-term human activity, including both forest harvesting and human-caused fires. As is clear from the diagram, over 40% of the regional Scots pine stands have been disturbed. By rate of disturbance, they are followed by fir and larch stands.

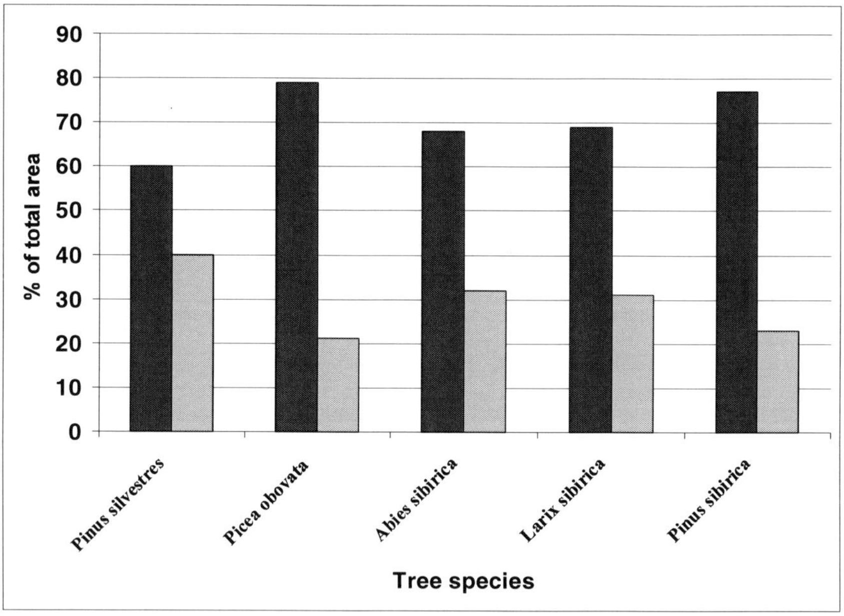

Figure 4. The ratio between undisturbed and forest areas in North Angara Region.

Structural Water Balance Changes Associated with Post-logging Forest Regeneration in West Sayan Mountains

Multiyear field hydrological observations conducted in mountain forests in West Sayan were based upon to assess the influence of post-logging forest recovery processes on the dynamics of the regional major water balance components. Clear cuts are known to markedly change catchments water balance. Particularly, precipitation interception by tree crowns and stand transpiration, which are among the major water balance components, are eliminated in clear-cut sites. As a result, runoff changes occur.

Investigations conducted on a sample logging site [4, 5] showed that total moisture evaporation on this site in the first post-logging year did not exceed half that in undisturbed forest. In years to follow, the amounts of moisture lost through transpiration and evaporation of intercepted precipitation on the sample site changed depending on vegetation rate of recovery [7, 8]. The difference in total moisture evaporation between sample logging site and undisturbed (control) forest site was insignificant eight years following logging. This characteristic tended to decrease on sample site compared to undisturbed forest in the twelfth post-logging year (Figure 5a).

In post-logging stands, total moisture evaporation amount and structure are determined by forest succession trajectories. While a post-logging birch and aspen stands exhibited a clear trend for total moisture evaporation to increase between 25 and 50 years of age, a fir stand showed maximum total moisture evaporation 35 years after logging, since when the characteristic tended to decrease (Figure 5b).

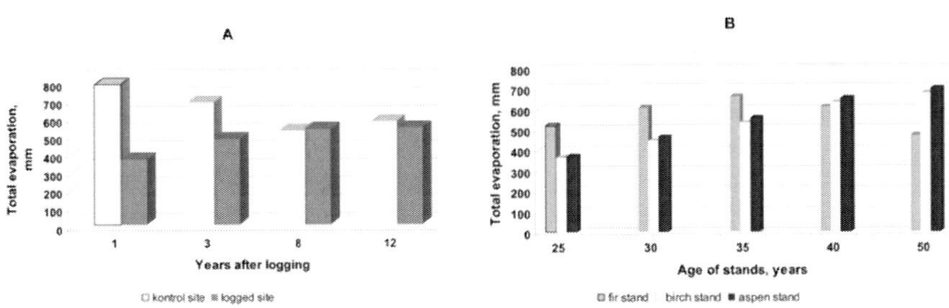

Figure 5. Total moisture evaporation on logged sites and in post-logging forest sands. Control site is Siberian pine stand, age160 years.

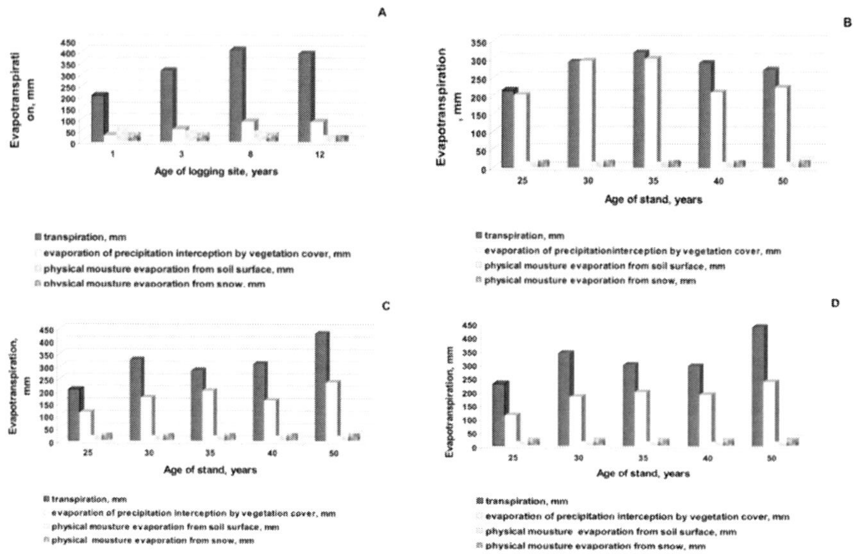

Figure 6. Structural changes of the total moisture evaporation during post-logging forest regeneration in West Sayan Mountains: logged sites (a); a fir stand (b); a birch stand (c) and an aspen stand (d).

Total moisture evaporation experiences structural changes during post-logging vegetation succession. The ratio of transpiration to intercepted precipitation evaporation differed considerably between post-logging conifer and deciduous young stands. Fifty years following logging, the young birch and aspen stands were found to loose twice the amount of moisture in transpiration compared to the young fir stand (Figure 6).

Since measuring the runoff for a relatively small location (a logged site or a sample forest site) *in situ* is problematic, it is usually determined using the water balance equation. As it is presumed that no changes of ground water and the amount of water contained in the aerated soil layer occur during a multiyear period of time [21, 24] this runoff is calculated as the difference between the total precipitation received by and moisture evaporated from a given location within a watershed. Runoff from fresh logging sites was higher to be almost twice that from undisturbed forest sites. A high rate of vegetation recovery on logged sites resulted in decreasing the runoff from these sites, and after eight years it approaches the runoff from undisturbed stands. Twelve years after logging, the runoff from logged sites began to increase and at the logging site age of 25 (aspen and birch stands) it became 30-40% higher that measured for undisturbed mature Siberian pine/fir stands. Runoff from 40-yr-old post-logging aspen and birch stands was equal to that from undisturbed mature dark-needled conifer stands and tended to decrease by the logging site age of 50. The dynamics of runoff in a post-logging fir stand differs from tendency of runoff at post-logging aspen and birch stands, since evapotranspiration in this stand was controlled by the stand structure and fir age.

The above observations allowed us to analyze dynamics of hydrological regimes of logged sites and to identify water balance structure for various post-logging forest regeneration trajectories in dark-needled conifer forest of chern taiga AVB in West Sayan. Water balance in 8-10-yr-old logging sites began to recover, which recovery was indicated by the fact that total moisture evaporation in these sites became the same as in undisturbed sites, with runoff growing more stable. The runoff in post-logging stands was controlled by stand species composition, age structure, and transpiring biomass amount.

Spatial Water Balance Patterns in River Catchments on the Northern West Sayan Macro Slope

Changes of stand structure and productivity caused by extensive principal felling over the past fifty years have influenced the hydrological parameters within each river catchment. Water balance changes identified for watershed sites disturbed by logging were determined to depend on forest succession characteristics. A landscape-hydrological approach based on a concept of close interaction between landscape and hydrological elements was used to estimate hydrological importance of vegetation communities found on different watersheds. The water balance parameters calculated for stands of the major and secondary woody species found in Kebezh and Taigish river basins allowed us to identify water balance structural changes in a number of altitudinal forest belts caused by human impacts on the northern macro slope of Western Sayan and estimate contributions of vegetation communities to Kebezh and Taigish river runoff characteristics.

According to our calculations, plant transpiration and evaporation of plant-intercepted liquid and solid precipitation are the major processes accounting for 80-95% of the total moisture evaporation in mountain and chern taiga AVBs. The ratio between these two water balance components differs considerably between undisturbed dark-needled conifer and secondary post-logging mixed conifer/deciduous stands.

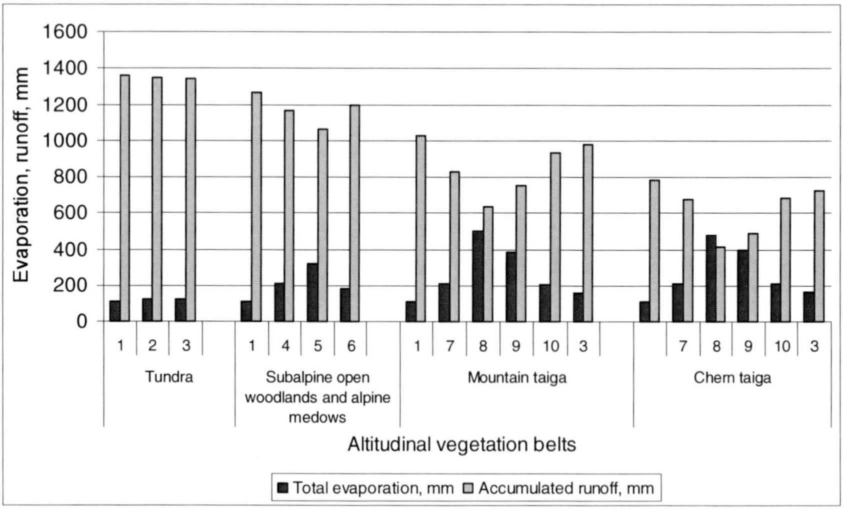

Figure 7. Total moisture evaporation / runoff ratio for landscapes of the northern West Sayan macro slope. 1 – bare rocks; 2 – dwarf birch; 3 – bog-grass communities; 4 – grass/small shrub communities; 5 – open dark-needled conifer woodlands; 6 – alpine meadows; 7 –open woodlands and shrub sites; 8 – dark-needled conifer forest; 9 – mixed secondary stands; 10 – logged sites.

Since runoff formation depends on ratios between landscape-specific water balance components, the estimated evapotranspiration values obtained for stands found in the river catchments of interest enabled indirect assessment of stand contribution to the runoff. The runoff from the vegetation communities identified in the satellite images was calculated as the difference between the total precipitation and total moisture evaporation.

The data obtained were analyzed to reveal that bogs and logged sites provide the biggest runoff in both AVBs. The runoff appeared to be 25-30% and 12-15% bigger from logging sites and post-logging conifer/deciduous stands compared to undisturbed sites (Figure 7).

The amount flow (volume of water) from each of landscape class and its percentage in the total runoff were calculated with an account it's area within each of the two catchments. Over 80% of Kebezh runoff is contributed by runoff from chern and mountain taiga forest stands accounting for almost 90% of the total catchments area, the respective values being 65% and 72% for Taigish river. While the runoff from logging sites markedly exceeds the runoff from undisturbed forest sites in amount, it does not contribute considerably to

Kebezh and Taigish channel runoff today, because the logged sites have become smaller in size.

The fact that Kebezh and Taigish catchments flows are mainly contributed by accumulated flow from West Sayan mountain forests indicates that these forests perform important water regulating functions. We used a water balance method to assess the proportion of precipitation contributing to accumulated flow in each AVB. The values obtained were 90%, 45-55%, and 55-70% for tundra-bare rock AVB, undisturbed Siberian pine/fir stands, and secondary post-logging stands, respectively. As is clear from the above percentages, secondary stands are an important catchments flow control. Although undisturbed and secondary post-logging stands became equal regarding evapotranspiration and runoff grew relatively stable 40-50 year after logging, the water balance structure characteristic of undisturbed Siberian pine stands did not recover fully in post-logging stands and, consequently, neither did their water regulating functions.

Changes of Channel Runoff of the Rivers within Angara Catchments Induced by Forest Harvesting

A number of studies of hydrological effects of extensive forest harvesting were conducted in Angara region [26] to reveal that logging sites induce the biggest effect on seasonal runoff structure. Changing catchments forest area is first of all reflected in water redistribution between runoff horizons. Wood extraction of 10-50% increases surface runoff by 10-20% and reduces ground (depth) runoff by the same percentage. Surface runoff increases by 2-7% in the case when fresh logging sites total 3-10% of middle-length channel catchments in southern taiga. Ground runoff exhibits a decrease of 10-20% where 20-65% of catchments forest was harvested.

Clear cuts induce changes of seasonal runoff structure. Annual runoff redistribution was found to be manifested in increasing surface runoff, and hence an increase in spring surface runoff proportion in annual runoff, under large-scale logging in catchments of North Angara region. This effect is more readily observable in catchments subject to large-scale clear cuts. Spring surface runoff is reported to increase by 5-7% in 4000-5000 km^2 catchments containing 1% fresh-cut sites. Increasing logging area to 35-40% results in increasing spring surface runoff by 10-20%. As for small rivers, post-logging spring runoff accounts for up to70-90% of the total annual runoff [26].

Human-caused alterations of soils and vegetation result in changing runoff genesis, hydro climatic conditions remaining the same. Our studies determined that these changes first occurred when the catchments of interest were covered by extensive forest harvesting, establishment of post-logging young secondary deciduous stands being another trigger of the changes [6, 31].

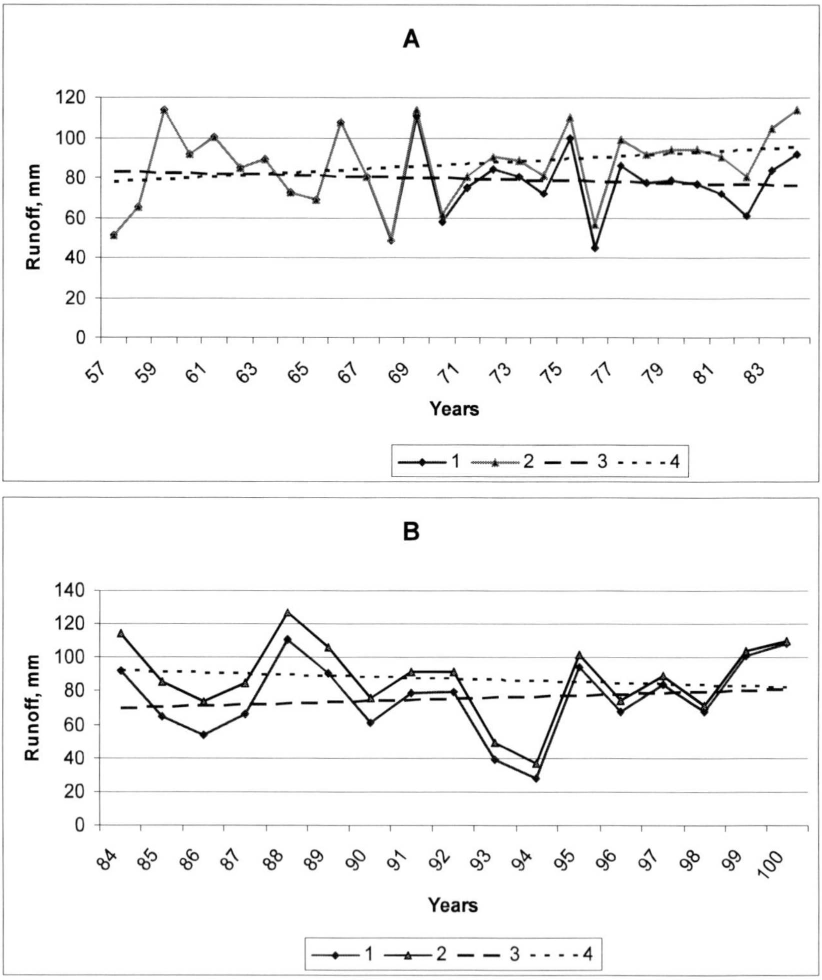

Figure 8. Mura river runoff changes resulted from forest logging and post-logging vegetation succession. A – time after forest logging; B – time of post-logging vegetation succession; 1 – actual (logging-affected) runoff; 2 – calculated (unaffected) runoff; 3 – trend of actual runoff; 4 – trend of calculated (unaffected) runoff.

The annual runoff of most rivers of North Angara region, where extensive forest harvesting was started in 1960s, is closely linked with the time factor. For three rivers (Karabula, Irkineyeva, and Mura), a twenty-year period beginning in early 1960s was identified when their annual runoff decreased with time most probably due to increasing fresh logging sites characterized by increasing wind activity and wintertime solid precipitation evaporation. The variation of channel runoff response to forest harvesting identified among the rivers of interest might be associated with differences in time of harvesting and post-logging vegetation succession trajectories including or excluding conifer woody species replacement by deciduous species.

A gradual runoff decrease (ca. 0.5-1.3mm annually) was found within a time period of about twenty years, between 1960s and 1980s, for Karabula, Irkineyeva, and Mura rivers. This decrease is presumably attributable to intense logging activities started in 1960s. On the resulting extensive clear cuts, wind activity and snow moisture evaporation increased. Thus, river runoff was anthropogenically reduced by about 10-20 mm during the first two decades of commercial logging in the Angara region. However, all rivers, except for Karabula, have tended to increase their runoffs since certain years (1975 for Taseyeva, 1984 for Irkineyeva and Mura, and 1986 for Chadobets). These years are approximately when young secondary forest stands began to expand on the 1950s-1960's clear cuts. The figure 8 shows Mura river runoff changes resulted from forest logging and post-logging vegetation succession. We believe the runoff increasing is connected with insignificant snow interception by tree crowns and capability of deciduous woody species to accumulate snow.

This is confirmed by the observed syneffect which suggests increasing snowfall importance to river runoff formation in time (for Taseyeva, Chadobets, and Irkineyeva rivers). An increase in runoff due to increasing snow accumulation in secondary young stands was estimated to range 1.5 to 3.0 mm annually. Thus, average annual river runoff increased by 20-40 mm during two decades since secondary young stand establishment on clear cuts as compared to before extensive forest harvesting.

The river runoff data we collected for the river basins of the North Angara region were based upon to develop equations (table 3) describing annual river runoff dependence on hydrological, climatic factors, and time since forest logging for.

Table 3. Runoff models of rivers in North Angara region

River	Equation	R^2	G	F
Taseyeva	$Y=-95,8+0,9*LnX_z*T_{75}+198,3*LnX_g*Y_p/(t_5+t_7+5)+1,22*X_g/(t_g+11)$	0,45	24,8	15,5
Chadobets	$Y=41,9+0,0054*X_z*T_{86}+0,23X_z*X_l/(t_5*t_7)$	0,50	26,2	20,8
Irkineyeva	$Y=1150,0+0,103*T_{84}*(X_z-11,5)+0,202*X_z*X_l/(t_5*t_7)-1,03*T_{60}-8,64*X_z$	0,51	25,8	9,3
Mura	$Y=91,0-1,3*T_{67}+1,3*T_{84}-5,3*t_7+7,3*t_9+1,8*t_9*t_g+3,25*X_g/(t_g+11)-0,35*X_l$	0,53	14,3	5,8
Karabula	$Y=58,2+0,121*X_z*X_l/(t_5*t_7)+0,54*t_9*(t_g+11)-0,5*T_{62}$	0,46	15,8	12,1

Y - annual runoff, mm; Y_p - previous year runoff, mm; X_z - precipitation as snowfall recorded by the nearest weather station (NWS), mm; X_g - annual precipitation recorded by the NWS, mm; t_5, t_7,t_9 - mean air temperature in May, July and September; t_g - average annual air temperature; X_l - precipitation as rainfall recorded by the NWS, mm; T_{60} = 60 prior to 1960, T_{60} = NN between 1960 and 1984, T_{60} = 84 after 1984; T_{62} = 62 prior to 1962, T_{62} = NN between 1962 and 1980, T_{62} = 80 after 1980; T_{67} = 67 prior to 1967, T_{67} = NN between 1967, 1984, T_{67} = 84 after 1984; T_{75} = 75 prior to 1975, T_{75} = NN between 1975 and 1999; T_{84} = 84 prior to 1984, T_{84} = NN between 1984 and 1999; T_{86} = 86 prior to 1986, T_{86} = NN between 1986 and 1999; NN = 19NN-1900;; R^2 - multiple determination coefficient; G - standard error; F - Fisher criterion.

Our analysis of the hydrological models developed for rivers of the Angara region revealed that their annual runoffs are significantly related with a number of hydro climatic parameters. All rivers exhibited an increase in annual runoff with increasing precipitation, while it was found to decrease with increasing July and, except for Mura river, May air temperatures. For two rivers, Karabula and Mura, increasing annual runoff appeared to be associated with an increase in September air temperature, and this fact indirectly indicates that the runoffs of these rivers are to a certain extent fed by permafrost water from seasonally frozen upper soil layers.

Our study revealed that hydrological effects of forest logging are determined by the background climatic parameters and logged site recovery characteristics. Under severe forest growing conditions making vegetation regeneration a much-time taking process, strong wind and high snow evaporation promoted on clear cuts induce river runoff reduction. As secondary deciduous woody species regenerate on logged sites, their snow holding capability increases and, as a result, runoff increases and even exceeds, in time, that on catchments not subject to forest harvesting.

Our comparison showed that West Sayan runoff and that calculated for North Angara Region differed in dynamics. This difference is largely due to the fact that the two areas are climatically different. Angara Region is in an extremely continental climate with severe winter and hot summer, whereas the northern macro slope of West Sayan Mountains is in a milder and wetter climate. According to a landscape-based concept of the hydrological role of forests, the hydrological effects of forests are determined by geographical conditions which control water cycling mechanisms [33]. Therefore, the same forest ecosystem may either contribute to stream channel flow, or become a soil moisture 'evaporator' depending on climatic and weather conditions.

In a milder and wetter climate characteristic forests are of a higher productivity and they act as a better moisture evaporator in winter as compared to open sites. There are two reasons for that: (1) decreasing nonproductive snow moisture evaporation on open sites, as moist and compact snow does not promote development of snow transport (it is known during snow storms snow evaporation increases exponentially with wind speed) and (2) increasing moist snow interception by the closed canopy of highly productive forest stands. In these conditions, a forest is a factor reducing snow pack. In a cold climate, the forest, on the contrary, acts as a snow moisture 'accumulator'.

From logged sites found in a mild climate runoff is higher than in forest sites. Such an increase occurs because certain water budget components (snow interception by forest canopy and stand transpiration) are omitted on a logged site. The difference of open vs. forest site moisture evaporation from snow surface is inconsiderable. This suggests that the water yield from logged sites increases due to increasing snow accumulation on such sites. A decrease in the water yield that follows is a result of increasing evapotranspiration of dense and high-productivity young stands developing on logged sites.

In continental Siberia, where relative humidity is relatively low in winter and logged site recovery takes a long time due to harsh forest growing conditions, post-logging hydrological regime changes are locally specific. On vast clear cuts, wind activity grows, in the first several years after logging, to promote snow storms and hence to increase snow moisture evaporation and to decrease snow pack. Under similar background climatic conditions, this results in decreasing water yield of catchments subjected to clear cutting. As forest stands gradually recover on such vast clear-cut sites, especially where the recovery occurs through tree species conversion, post-logging woody vegetation regains, and even increases, its snow accumulation function and the water yield of the sites increases.

Our studies show that, in Siberia, clear cutting may result either in increasing or in decreasing water yield, depending on forest cover structure and climatic conditions. Water yield trends are considerably controlled by inter annual precipitation changes. In Lake Baikal basin, for example, in high-precipitation years forest harvesting increases annual water yield by 25-30%, depending on the area logged and the precipitation amount in a given year, whereas in low-precipitation years forest harvesting may either have no effect on, or reduce annual runoff [25, 36].

CONCLUSION

Study areas, West Sayan and North Angara region, are nowadays a mosaic of forest regeneration sites including both post-human and post-fire regeneration patterns. These forest cover changes influence hydrological regime, particularly the total moisture evaporation and runoff. The rate of regeneration in logged sites controls moisture amount spent on transpiration and intercepted precipitation evaporation. Water balance calculated for secondary forest stands showed that evaporation structure and total amount significantly depend on forest succession trajectory. The runoff in post-logging stands was controlled by stand species composition, age structure, and transpiring biomass amount. Secondary stands are an important runoff control.

Hydrological effects of forest logging are determined by the background climatic parameters. It is connected with a fact that in a boreal zone snow cover have especially important hydrological role which consists in essential difference of snow moisture balance on forest and open sites. In West Sayan with more mild and humid climate snow moisture evaporation was more intensive in the forest than in open places, as opposed to the trend found for northern Angara study sites. This can be attributed to the fact that extremely continental climate with high wind activity characteristic of Angara region enhances evaporation in open places and accordingly exceeds the evaporation in the forest. This explains the occurrence of contradicting views of hydrological importance of forests in different geographical regions.

Depending on weather, climatic, landscape conditions and also on structure of forests cover, forests logging can act both as the store of a moisture and the source of river runoff and as the factor resulting in increase in total evaporation and reduction of river runoff. The method used to analyze annual river runoff genesis allowed to date the onset of anthropogenic disturbances and assess their level. Moreover, changes of river runoff formation genesis can be based upon in land use decision-making. River runoff reflects all catchments changes, effects of which can be opposite in sign and thus compensate for each other.

REFERENCES

[1] Andreyashkina, N. I. & Gorchakovskiy, P. L. (1972). Productivity of small shrub and grassy associations of forest- tundra and methods of its assessment. *J. Ekologia. 3: 23-34*. (In Russian).

[2] Bazilevich, N. I. (1993). *Biological productivity of ecosystems of Northern Eurasia*. Moscow. Nauka. (In Russian).

[3] Bosch, J. M. & Hewlett, J. D. (1982). A review of catchment experiment to determine the effect of vegetation changes on eater yield and evapotranspiration. *Journal of Hydrology, 55: 3-23.* (In Russian).

[4] Burenina, T. A. (1981). Water balance structure dynamics in logged site of West Sayan. In V. V. Protopopov (Eds). *Actual problems of Siberian forest research*. (19-20). Krasnoyarsk, Publ. House of Institute of Forest & Wood SB, USSR Academy of Science (In Russian).

[5] Burenina, T. A. (1982). Dynamics of hydrological regime restoration at the experimental logged plot in West Sayan. In V. V. Protopopov (Eds). *The Environmental role of Siberian forest ecosystems.* (106-114). Krasnoyarsk, Institute of Forest & Wood SB, USSR Academy of Sciences. (In Russian).

[6] Burenina, T. A. & Onuchin, A. A. (1999). Runoff dynamics at North Angara region watersheds. *J. Geography and Natural Resources, 2, 2-46*. (In Russian).

[7] Burenina, T. A., Ovchinnikova, N. F. & Fedotova, E. V. 2011. Water balance dynamics in logged sites of chern taiga in West Sayan. *J. Geography and Natural Resources. 1,* 92-100 (In Russian).

[8] Burenina, N. A., Fedotova, E. V. & Ovchinnikova, N. F. (2012). Change in the structure of the hydrological cycle in connection with the age and recovery dynamics of forest ecosystems. *Sibirskii Ekologicheskii Zhurnal. 3,* 435–445 (In Russian).

[9] Yermolenko, P. M. (1981). Forming of mixed young grows composition at clear cuttings in chern zone of West Sayan. In N. P. Polikarpov (Eds),

Forest regeneration and productivity of stands (53-70). Novosibirsk, Nauka (In Russian).

[10] FAO. (2006). Global Forest Resources assessment 2005 – progress towards sustainable forest management. FAO Forestry Paper. Rome.

[11] Farber, S. K., Sokolov, V. A., Kazimov, S. A. (1995). Dynamics of regeneration at logged sites of pine forests of Priangarye. *J. Lesnay promishlennost. № 3,*pp. 26-35 (In Russian).

[12] Gorbatenko, V. M. & Onuchin, A. A. (1976). Studying the components of forest biogeocenoses of Siberia. In V. V. Protopopov (Eds). *Studying the components of forest biogeocenoses of Siberia.* (129-132). Krasnoyarsk, Institute of Forest & Wood, SB, USSR Academy of Sciences, 1976 (In Russian).

[13] Gorbatenko, V. M., Kozlova, L. N., Onuchin, A. A. (1977). Transformation of water balance elements by dark- and light coniferous forest plant formations at the Khamar- Daban ridge under different a forestation degree of catchment basins. In V.V. Protopopov (Eds). *Ecological forest impact on environment.* (116-146). Krasnoyarsk, Institute of Forest& Wood, SB, USSR Academy of Sciences. (In Russian).

[14] Grip, H., Fritsch, J. M. & Bruijnzeel, L. A. (2005). Soil and water impacts during forest conversion and stabilization to new land use. In M. Bonell & L. A. Bruijnzeel (Eds). The Forests, water and people in the humid tropics. UNESCO International Hydrology Series. (561–589). Cambridge, UK, Cambridge University Press.

[15] Hamilton, L. S. (1988) The recent Bangladesh flood disaster not caused by deforestation alone. *Environmental Conservation. 15(4),* 369-370.

[16] Hamilton, L. S. & Pearce, A. J. (1987). What are the soil and water benefits of planting trees in developing country watersheds? In D. D. Southgate and J. F. Disinger (Eds). *The Sustainable resource development in the third world.*(39-58). Boulder, Colorado, USA, West views Press.

[17] Hewlett, J. D. (1982). Forests and floods in the light of recent investigation. In *Proceedings, Canadian Hydrological Symposium.* (543–560). Fredericton, New Brunswick, Canada, 15 June 1982. Ottawa, National Research Council.

[18] Hydrology Guide. (1986). 7, 5-8, Gidrometeoizdat Publishing, Russia.

[19] Kozlova, L. N. (1980). Expense of water for transpiration by southern taiga plants the Kas-Yenisei plain. In V. V. Protopopov (Eds) The

Protective role of Siberian forests. (46-61). Krasnoyarsk, Institute of Forest and Wood, SB, USSR Academy of Sciences (In Russian).

[20] Kozlova, L. N. (1988). Transpiration in aspen-birch stands of Krasnoyarsk forest steppe. *J. Lesnoye khozyaistvo. 7.* 33-34. (In Russian).

[21] Krestovskiy, O. I. (1984). Logging and forestation impact on water-yield in southern and middle taiga of ETU. *J. Water resources, 5*, 125-135. (In Russian).

[22] Krestovskiy, O. I. (1986). *Logging and forestation impact on water-yield of rivers.* Leningrad, Hydrometeoizdat. (In Russian).

[23] Kuzmichev, V. V, Ovchinnikova, N. F., & Yermolenko, P. M. (2002). Restoring dynamics of dark-needled forests at clear cuttings in West Sayan. *J. Lesnoye khozyaistvo, 6,* 22-24. (In Russian).

[24] Lebedev, A. V. (1982). *Hydrological role of Siberian mountain forests.* Novosibirsk. Nauka. Siberian Branch. (In Russian).

[25] Lebedev, A.V., Gorbatenko V. M, Krasnoshchekov, Y. N. (1979). *Environmental role of forests in the Baikal basin.* Novosibirsk. Nauka. Siberian Branch. (In Russian).

[26] Lebedev, A. V., Spiridonov, B. S., Moreva, L. S. et al. (1984). *Transformation of environmental factors by forest ecosystems.* Krasnoyarsk, Publ. House of Institute of Forest & Wood, SB, USSR Academy of Sciences (In Russian).

[27] Nisbet, T. R. & McKay, H. (2000). Sustainable forestry and the protection of water in Great Britain. In *Proceedings, International Expert Meeting on Forests and Water.* (101–112). Shiga, Japan. Tokyo, International Forestry Cooperation Office, Forestry Agency.

[28] Onuchin, A. A. (1985). Biomass of crowns and needle of Siberian pine and fir tree stands of Khamar- Daban. In V. V. Protopopov (Eds.). *The Structure, growth and inventory of tree stands.* (32-41). Krasnoyarsk, Publ. House of Institute of Forest & Wood SB, USSR Academy of Sciences. (In Russian).

[29] Onuchin, A. A. (2009). Hydrological role of forest ecosystems of boreal zone. Ecological and Geographical Aspects of Forest Forming Process. In *the Proceedings of the Int. Conference 23-25 September 2009.* (11-14). Krasnoyarsk, V. N. Sukachev Institute of Forest, Russian Academy of Sciences (In Russian).

[30] Onuchin, A. A., & Borisov, A. N. (1984). Experience in inventory of pine tree stands phytomass. *J. Lesovedenie. 6,* 66-71. (In Russian).

[31] Onuchin, A. A., & Burenina, T. A. (2000). Anthropogenic dynamics of soil preventive and water protective functions of Siberian mountain taiga forests. *J. Lesovedenie. 1,* 3-11. (In Russian).

[32] Onuchin, A. A., & Burenina, T. A. (2008). Hydrological role of forests in Siberia. In A.P. Prescott (Ads.) *Trends in Water Resources Research.* (67-92). Nova science publishers.

[33] Onuchin, A. A., & Burenin, T. A. (2011). Conception approach in assessment of hydrological role of boreal forests. In *Proceedings science conference of Russian geographical society.* (34-36). Krasnoyarsk, Russian geographical society. (In Russian).

[34] Onuchin, A. A., & Spitsyna, N. T. (1995). Regularities of varying needle mass in coniferous stands. *J. Lesovedenie. 5*, 45-57. (In Russian).

[35] Polikarpov, N. P. (1970). Problems of Forestry. In A. B. Zhukov (Eds.). *Complex Studies in the Mountain Forests of Western Sayan.* Vol. 1. (3-16). Krasnoyarsk, Institute of Forest SB RAS USSR (In Russian).

[36] Protopopov, V. V., Lebedev, A. V., Bizyukin, V. V., et al. (1991). Water protective role of forests of the lake Baikal basin. In *Proceedings, Science conference on problems of water resources of the Far Eastern economic region and Zabaikalie.* (298-310). S- Petersburg, Hydrometeoizdat. (In Russian).

[37] Rao, Y. S. (1988). Flash floods in southern Thailand. *Tiger Paper. 154,* 1–2.

[38] Resources of surface waters of the USSR. (1973). Vol. 16. (1-3). Leningrad, Hydrometeoizdat. (In Russian).

[39] Riano, D., Chuvieco, E., Salas, J., & Aguado, I. (2003). Assessment of Different Topographic Corrections in Landsat-TM Data for Mapping Vegetation Types. *IEEE Transactions On Geoscience And Remote Sensing, 41 (5), 1056-1061.*

[40] Richards, J. A. (1993). *Remote Sensing. Digital Image Analyst.* Second Edition: Springer-Verlag.

[41] Sokolov, V. A., Atkin, A. S., Farber, S. K. et al. 1994. *Structure and dynamics of boreal forest.* Novosibirsk, Nauka. Siberian Branch. 168 pp.

[42] Sun, G. (2008). Forest hydrology in China: Introduction to the featured collection. *Journal of the American Water Resources Association. 44 (5),* 1073-1075.

[43] The USSR Climate Guide. (1969). Vol. 2-3, 21. Leningrad, Gidrometeoizdat Publishing, Russia.

[44] Wang, Y. (2008). Water-yield reduction after a forestation and related processes in the semiarid Liupan Mountains, northwest China. *Journal of the American Water Resources Association, 44 (5),* 1086-1097.

[45] Wei, X., Zhou, X. F., & Wang, C. K. (2003). Impacts of the temperate forests on hydrology, Northeast of China. *Forest Chronicle. 79,* 297-300.

[46] Usol'tsev, V. A. (2002). *Phytomass of forests of northern Eurasia: data base and geography.* Yekaterinburg, Ural Branch of Russian Academy of science (In Russian).

[47] Hydro meteorological centre of Krasnoyarsk kray. Available from http://meteo.krasnoyarsk.ru/

INDEX

D

E

F

G

H

I

J

K

L

M

N

O

P

W